Writing for Communication in Science and Medicine

Barbara S. Lynch, Ph.D.

Writer and Consultant, Washington, D.C.; and Faculty Member, Department of Health Education, University of Maryland

Charles F. Chapman

Director, Editorial Office, Louisiana State University School of Medicine, New Orleans

VNR VAN NOSTRAND REINHOLD COMPANY

NEW YORK CINCINNATI ATLANTA DALLAS SAN FRANCISCO
LONDON TORONTO MELBOURNE

Van Nostrand Reinhold Company Regional Offices:
New York Cincinnati Atlanta Dallas San Francisco

Van Nostrand Reinhold Company International Offices:
London Toronto Melbourne

Library of Congress Catalog Card Number: 78-18893
ISBN: 0-442-24959-4

Manufactured in the United States of America

Published by Van Nostrand Reinhold Company
135 West 50th Street, New York, N.Y. 10020

Published simultaneously in Canada by Van Nostrand Reinhold Ltd.

15 14 13 12 11 10 9 8 7 6 5 4 3 2 1

Library of Congress Cataloging in Publication Data

Lynch, Barbara S.

Writing for communication in science and medicine.

Includes index.

1. Technical writing. I. Chapman, Charles Frederick, 1929- joint author. II. Title.

T11. L86 808'.066'5021 78-18893
ISBN 0-442-24959-4

List of Dosages

Introducing Lapsus. Of Latin ancestry, *Lapsus* means fault, error, fall, or slide, and is cousin to the meanings of glide and even sleep. Known by several full names—Lapsus Linguae (slip of the tongue), Lapsus Calami (slip of the pen), Lapsus Memoriae (a slip of the memory)—he serves well as a serpentification of problems in science writing, especially as his great uncles were twain with Aesculapius, Mercury, and Medusa's hair style. At the end of him-limb is the measured foot, perhaps anthropomorphosizing him somewhat. The staff, by the way, was unnamed until now: *Logos*, Greek for word, reason, speech, or account. Course with us and Lapsus Logos through the book to put some charge in your coil, and sidewind upward on the sturdy shaft of reason.

Preface

Our style throughout the book is light and casual for a serious purpose; we want you to enjoy your reading so that you will enjoy learning about writing. Also, we hope by example to show the richness of words and of their sequences of associations, and how structural patterns in sentences have meaning. Just as it is easier to find fault in others than in ourselves, it is often easier to spot patterns, or the forest in the trees, in materials or styles unfamiliar to us. If the science writer can learn some fundamentals about nonscientific language, he or she* may then be able to spot these same principles in science writing. For this reason, some of the material in the text is not about science.

Our hypothesis is that once you understand words, sentences, the logical patterns behind them, and their symbolic workings and implications, you will easily perceive the patterns of words and their skillful expression. You will more easily spot your own errors and, by applying your mind to the arrangement of words, you will achieve not only quality in writing but also an enjoyment in writing that rote exercises can never give you.

To this end, *Writing for Communication in Science and Medicine* presents discussions of theory and assignments or games for the reader that, if he cooperates with them, will lead him to certain concepts about language usage. Examples and exercises are also provided for each principle, with correct answers in the text or Appendix. Please follow each stage of the text and do not jump ahead to exercises or answers; they are practice for the principles that are presented in the text, and are important as an elucidation of those principles. If you carefully follow the text's program, you may be assured that you can indeed be comprehended.

Early in the text, a few basic principles of language and communication are discussed so that the problems of communication and the

*When we asked God to enlighten us about how to handle this current gender dilemma in our book, she answered, "Go traditional." (What would you recommend?)

reasons for writing conventions will become clear. Understanding the "theory" or what is behind the mind's door is the important and essential preface to learning how to write, so that as author you will be at ease, knowing that you know why you are doing what you do. The unknown is intimidating, but the familiar can be comfortable. Except for the few for whom language comes intuitively, writers, especially science writers, need to understand their writing materials—words—and *why* those materials habitually form bonds with one another, even as the scientist or clinician must understand biochemical and physiological patterns before he can analyze experimental data or make a diagnosis.

The text describes the type of language and writing that is appropriate to the sciences. Therefore, when we characterize good science writing as being direct, pruned, and logical, we are not implying that this is the only kind of good writing. Writing for other purposes—historical insights, poetic lie-sense, editorializing, irony—may benefit from embellishment and illogic.

Fortunately for the scientist, whether he knows it or not, nothing in the avocation of writing is foreign to him. He may think various elements are foreign or that he must learn a new "discipline," but this unfortunate perception occurs only because subjects and skills are typically taught as distinct subjects. Actually, the very structures and tools of thought used in science are also basic to writing. Just as science seeks objectivity through systematic formulation and testing of thought, so, too, does writing seek communication through systematic formulation of language and reproducibility of thought.

Part I will lead you into the basic questions: Why does a scientist write? What are the mental blocks to writing and how can they be knocked down? What tricks or habits can facilitate writing? How can you critically appraise your own writing? What is communication and what are the tools of communication?

Selection and appropriate use of words in science writing are the focus and emphasis for the four chapters of Part II. The reader, through examples, exercises, and the active practice of writing, apprehends and appreciates the functions of words, individually and in combinations.

Part III explains how words form patterns essential to the communication of ideas. The morphology and functions of words are the framework for sentence syntax. As ideas move from words to sentences, sentences aggregate progressively to build paragraphs and fuller concepts.

The form and format of the science article undergoes scrutiny in Part IV. The dispassionate forces of economics have burst the sanctuaries of tradition and the test of time, causing questions to rise on the value of the tried and true forms, but basic advantages persist in certain formal elements.

The two chapters of Part V guide the reader toward actually publishing his manuscript. How will *you* edit your manuscript, and how will *you* prepare it for the journal you have selected?

Part VI is the book's handy shelf of tools you might find advantageous in developing your articles and your image as a writer: stylebooks, writing handbooks, dictionaries; a list of peeves, peevies, and other clarifications (in the Glossary); a commentary on what to expect after publication; and answers to the Dosages in the text (in the Appendix).

Here, then, is a six-part pack for you, reader, unto your becoming you, writer.

Do (and) enjoy!

BARBARA LYNCH
CHUCK CHAPMAN

Contents

Writing for Communication in Science and Medicine

PART I. COMMUNICATING: MINDING THE STORE OF WORDS

1. Why Write?

When Dr. Richard M. Paddison, Head of the Department of Neurology at Louisiana State University Medical Center, was recruited for the position of Dean of the Medical School, he emphatically refused with the explanation that "Deans are just like hemorrhoids–they're a pain in the ass, they're always hanging around where they're not wanted, and when they get to be a real problem, they're chopped off." Thereby hangs a tale.

Somewhere in that natural history and prognosis of hemorrhoids is the wisdom that could be a guiding philosophy for science writing. Certainly, for many scientists, it is descriptive of their attitude toward scientific articles and the obligations to write them. Perhaps we can ease the specific pain of the sore word by applying the same therapy.

If faulty writing can be considered a disease syndrome, then its symptoms must be recognized and therapy applied. This book is a type of elixir, Preparation W, which, if taken carefully according to the prescribed dosages, would cure the writer of more than his writer's cramps. He will know how to excise malignant sentences, swollen and sore phrases, and dead words from his manuscript. He will move his scientific thoughts and pen more comfortably and boldly in his sensitive situation of communicating objective truth. But for all to heal well, he will have to carefully and thoroughly take each dosage of the stepped therapy. To get the most benefit from this book, you are requested to participate in all of the suggested writing exercises because we believe that the best way to learn to write is to use the language, but to use it in ways you have not necessarily done before.

The writing exercises suggested in this book are called "Dosages" because each one is a different prescription directed toward achieving a particular concept or writing skill. The Dosages follow progressively, each prescribed with the assumption that the previous Dosage has been taken and has effected its therapeutic purpose. Because they are progressive, you, the reader, are asked to record each Dosage in a singular-to-purpose notebook so that you will have a permanent record of your course and can review your writing recovery progress. The notebook, which will be referred to as a *journal*, should have a special physical identity—color, page size, ruled or unruled—and if possible it should be bound (not loose-leaf or spiral). The journal should eventually take on an identity of its own, but also become a reflection of yourself with a mem-brain of ideas you ordinarily would have forgotten. Give it a name: Kaleidoscope, The Book, Expository, In Quest, I-Go Ego, Thought-Put Triumph, Harvest Time, Fill in the Blank, or ____________________.

Each Dosage should be numbered and dated as it is entered into the journal. Each journal entry will take only fifteen to twenty minutes, sometimes less, though we recommend writing for at least fifteen minutes every day to establish a writing habit. Incidentally, we have tried to make the Dosages fun, cherry flavored as it were. We hope you enjoy the regimen.

But before approaching the symbols and patterns of communication, scalpel in hand, we need to review the science writer's purpose in writing—his motivation—and this will determine the semantic philosophy he best adopt. Motivation, after all, is derived from Latin words meaning "to move," to "get it going." In the vernacular, this might translate, "getting off one's ass," and is certainly apropos in the cosmic systems of rewards and balances most science writers are painfully pressured by.

Let us begin by talking about YOU, WRITER.

MOTIVATION

Writing can be pleasurable, satisfying, and even fulfilling. The composite of your experience is unique to you; your knowing, seeing, sensing, and perceiving are peculiar to you. Use of that uniqueness commingled with your personal philosophical viewpoints in reporting scientific observations, making comparisons, drawing analogies, or stating conclusions can be the fun of meeting a challenge, providing a service, pedestalling the ego, or pursuing a gain. Satisfaction in writing may mean the mere publication of a piece, the payment for such publication, the notice by peers, or the advance of science.

Fulfillment in writing is truly the mature form of the pleasure-

Fig. 1-1. One's greatest need is to get off, or be lifted by circumstances from, one's *dupa*.

satisfaction extension. Writing leads to more writing, and its perpetuity can be something hungered for once you have the first jot or tittle under title. The wonder of it is in having a thought roll around in the mind in seeming search of concepts with which it can become related, sometimes cleverly, often unusually, usually offhandedly. The roll-around of ideas in your mind has the roulette quality of passing many slots in which the specific ideas may become juxtaposed to a concept. The more rolls you take, the more likely you are to land on something that pays off. And the payoff is in the excitement of insight, discovery, or the procedure thereto. A writer may have one article published, or two hundred, but, as in scientific research and medical practice, fulfillment comes as much with the persistent habit of writing as with the product. As self takes on the identity of a writer, the sphere of experience is only half-filled without the activity of writing and the experimenting roulette mind yearns for the game. A few scientists have found this game of fun and energy in writing; if you have not, we hope to lead you into addiction.

EVIL'S ROOT BREWED

Meanwhile, other rewards are possible. Money, that which everyone always has not enough of, may or may not be a reward for the writer, depending on the outlet sought and the steadfastness of one's search. Perhaps (aha!) that is why the Fourth Estate is called *the press*. If money is your motive, you must first be aware of those publishers who will pay for your thoughts. In the spirit of paradox, the system

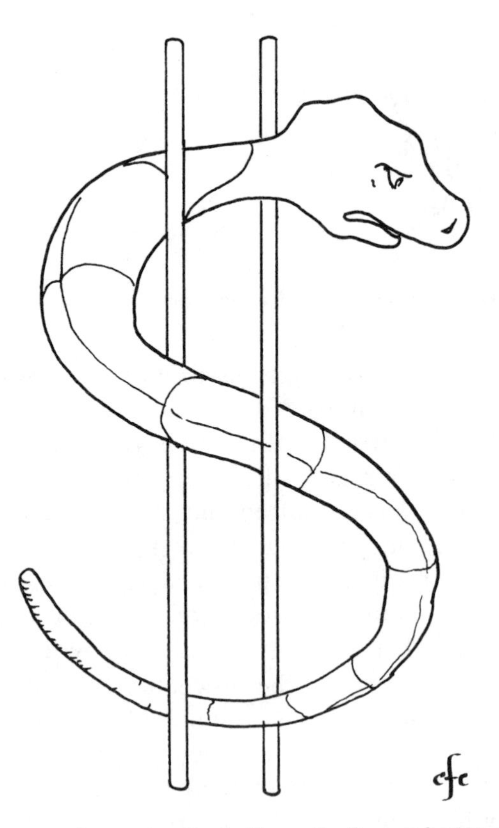

Fig. 1-2. Sometimes one becomes entangled in motivations, as Lapsus shows, aided by rod and staff.

does not always pay for the stuff you write. Many science and medical journals historically have considered that the publication of an article means it is worthy of dissemination, and that *that* should be reward enough for the author. Other journals, amending the historical viewpoint, actually have the author pay for publication of his message, an obligation called page charges or author fees. Sometimes the charges reflect the publisher's need to limit the cost for each article, and the author is required to pay only for the number of pages, illustrations, or use of color beyond the prescribed minimums. An incredible extension of the paradox is that the publisher in such cases customarily holds the copyright exclusively. Your work becomes his, and you pay him for the privilege.

Some publishers do pay, of course, but they are those whose products constitute "professional writing," a deliberately undefinable term. Often the readership is the lay audience. Ordinarily, the professional writer evolves from a simpler identity that is led by chance, others, and self into the professional genre. Probably a scientist's best bet would be to have other means of self-support while he is evolving his "professional" identity.

I-ING THE POSSIBILITIES

Beyond the dō (dough-money) and re (ray-illumination) is "me," or ego, needing spine-rubbing, achievement, recognition, and many unknowns. Whereas one can put the finger on most other rewards, the polish given to the ego can only be presumed. Fame, then, is the name of the ego game. In writing a piece you become *author*; in having it printed and disseminated, you become *published* (whence the question, "Are you published?"); in having it read you become *authority*. Please be endeared by the knowledge that authority begins with *author*, if not *author-it-*(wh)y? Why, for fame, of course.

Fame is not always an easy cloak to wear. The acquisition of fame provides coincident pressure to become familiar with business, finance, and locomotion, rewards unto themselves for many. Those with any level of fame are presumed to know many things; how to give a speech, how to be in three and a half places at one time; who the other authorities are in, around, and outside the field; how to pay for a trip to Peoria, get there by noon from home, and return to Palo Alto by midnight; and what hypertension really is.

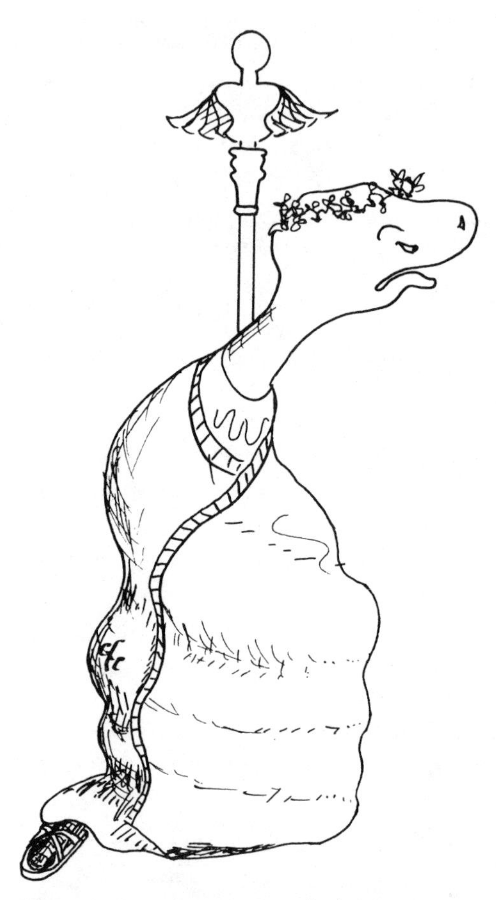

Fig. 1-3. Fame is not an easy cloak to wear . . .

PRINTSOME, TEND AND SEE

Immortality is an extension of fame into eternity and seems hardly achievable, but probably the best assurance that a representation of the self can be extended beyond the unto-dust is leaving in words the thought-images that now arise out of the dust. For some, then, the motivation for writing is the wonderful gift of self to the world. We are pleased to be communicated with by Sigmund Freud, who through his writings, tells us, "The problem of the nature of the world without regard to our precipient mental apparatus is an empty abstraction, devoid of practical interest. Science is no illusion. But it would be an illusion to suppose that what science cannot give us we can get elsewhere." As he is not here to reveal his true motivation in having written that, we cannot know for certain what his motive or meaning may have been, but each writer can certainly appreciate a personal extension of Freud or self in communications media, especially that of writing.

TIDE BENEATH THE NOSE

More down to earth and among the great wonders of academic life is a syndrome that has come to be called "publish or perish," although in Louisiana the term is excused as "publisher parish." Lamentations have been voiced on the demand by universities and scientific institutions that faculty members write, submit, and publish articles, monographs, and books. The true lament is that the pressure to publish is artificial, forcing academicians to pursue the goal of publication rather than that of satisfying specific needs and services. The defense would argue that many persons–among them academicians–require external pressure to produce the communications needed by others, that the pressure forces them to continue doing research that in turn produces excellence. Institutions themselves are reputed by the publications and authors they engender, and reputation means grant money for more research for more publications. Aggrandizement is reward, whether for person or place.

As a motive for writing, publish or perish may be less admirable than others, but it is decidedly evolutionary in the survival of the fittest. If this is your reason for writing, you are eager, if not anxious, to get those articles published. You might be somewhat resentful of the pressures and hence find the writing chore more tedious than need be (especially when it is sunny and windy and you could be sailing, or if it is sunny and not windy and you could be playing tennis), but you do like seeing that curriculum vitae lengthen, and you do enjoy the rewards of maintaining your position and of promotion. Hopefully, during the struggle for survival, the striving for excellence or the satisfaction in writing will accompany you.

HATFUL OF HEED

At any rate, to fitly survive, you will need to stay avant-garde in your field, and education is itself a rewarding motivation. Nowhere is the need to know greater than when one begins to write something, about which he is supposed to be an expert. Is each of an author's statements true? Has he considered the implications? Why has he not considered Dr. Schmouzer's latest finding on the confound of zero? Writing forces one to read, discuss, research, ponder, try–in effect, to use the grits, grist, and gristle we call education. Writers, by implication and application, are educated; pursuing full knowledge on a subject, they appreciate that each "known" unveils many unknowns, which in turn require knowing. Inasmuch as science is knowledge, science writers must be knowledgeable and knowledge is its own reward (ever enough?).

The fact is, your knowledge and discoveries are sought after. Each day, you doubtless have noted, the forces of journalism, media, intellectual curiosity, and interest in food and sex create a gigantic vacuum of and for facts and fables, opinions and options, traces and trivia, and news and merest clues thereof. Education, conditioning, society, and we individually provide the pressures for satisfying, if never fully gratifying, the need to know. Each week hundreds of magazines are printed, each month hundreds more, and each year thousands–thousands, not even including the plump monographs, books, special reports, newsletters, and enigmatic reap-prints. Somehow the needs of the vacuum are temporarily met: interviewing, "going to the literature," counting cases, hyper-theorizing a hypothesis. Facts pop from the instrument readings, an apt question is put to an "authority," the mind plays with a theme and a "discovery" is evoked, or two friends meet in a "have you heard?" session and each contributes to the refinement of the truth. The reality of the vacuum for information and the active pumping force of your own mind ought to be used to advantage. Write your thoughts and findings down, find the proper vacuum hose, and your reward will be in having satisfied someone's curiosity, as well as your own.

Now, suppose–and we will call this an ideal motivation for science writing–suppose you feel you have a message specifically for the scientific community. In your research or in your clinical practice you have come upon data or have synthesized experience that you think other scientists will benefit in knowing. You want what you have learned to be applied clinically, or repeated experimentally. You want not only to be published, but also to be *read*. Excellent, and this is the assumption we will make in the course of this book. But now the concept, motivation, and even the product of science writing assume a special complexion that needs a special type of scrutiny for its analysis. Because now, if you are writing for the purpose of being read, the urgent necessity is to write *well*. Therefore, we will spend a few pages on reconnaissance of the journal field and on what will motivate you to strive for perfection in this world if not in the next.

2. Who Needs It, Who Reads It?

When a court rules against a physician who did not know of the possible harmful side effects of a drug or other forms of therapy, the enraged public may be calmed, but fellow physicians become more anxious. How can they keep up with all the important literature? Now that many insurance companies are refusing to carry malpractice policies and states are beginning to require specified hours of continuing medical education each year for physicians to maintain their licenses, the physician wonders how he can learn enough to remain in practice and out of court. Among the proffered reasons, such as fatigue and lack of time, for his inability to keep up with scientific advances, the physician objects that too much is written and too little of it is read easily. That a patient might not be cured, or might even be harmed or killed because a physician just could not face reading another abstruse, boring journal article is an appalling threat. If all science articles were clear and came quickly to the point, what a lighter prospect doing one's homework would be, how much more would get read, and how many more patients would be helped. The exigencies of the patient-physician situation require that *medical* articles be well written; the need for information exchange between the medical and pure sciences, between clinicians and researchers, requires that all *science* articles be well written.

Coincident with a writer's idealistic or altruistic motive to write well so that his message will serve humanity is a practical motivation —fear of rejection. According to Mel Brooks's 2,000-year-old man, all social customs originated out of fear—even singing, the handshake, and dancing. When you were in trouble in the old days, explains the double-millenniumian, you had to sing it out so people would notice and help. If you made your plea in ordinary language, folks would pass on by even though your foot was foreleg deep in the mouth of a lion. You shook another's hand to be sure that he did not have a rock in it, and at social gatherings folks danced so they would know what the hands and feet of one another were doing.

Maybe we write out of fear of being alone with our thoughts and maybe we strive to write *well* out of fear of having our thoughts hurled back at us like a stone, jostling us out of our old platitudinous sticks-and-stones contentment. There is a sound basis for fearing that an article will be rejected. Journals are becoming more and more particular about what they will accept. Unlike in 1679, with publication of Nicolas de Blegny's *Nouvelles Descouvertes sur toutes les parties de la médecine*, considered the first medical periodical in the vernacular (lasting only until 1681), journals no longer lack materials; to the contrary, publisher parish is swamped with them. And it costs more to print them today. So publishers can afford to be choosey. You are the one perishing; they are publishing. They now must be strict about both content and style of writing; after all, the prestige of the journal depends on both.

Poorly written articles might be accepted by less prestigious journals—but then they are less likely to be read (and your purpose *is* to be read)—and you, your work, and its clinical potential will pass unnoticed. Do your own survey: how many journals do you read and how many of the articles in those journals do you read? Consider the information in the survey reported in *Medical Economics* (March 5, 1973):

Don't let rejection slips discourage you

To prove that they shouldn't, Dr. John G. Deaton compiled this list of 17 leading journals ranked by the ratio of unsolicited manuscripts rejected to those received in one year.

JOURNAL	MANUSCRIPTS RECEIVED	PERCENTAGE REJECTED
New England Journal of Medicine	2,160	83%
Science	6,100	77
Journal of the American Medical Association	3,150	75
Annals of Internal Medicine	1,198	75
Archives of General Psychiatry	525	64
American Journal of Medicine	500	60
American Journal of Psychiatry	838	54
Journal of Pediatrics	854	53

JOURNAL	MANUSCRIPTS RECEIVED	PERCENTAGE REJECTED
American Journal of Diseases of Children	552	49
Journal of Clinical Investigation	824	48
Radiology	1,004	45
Journal of Surgical Research	196	42
American Journal of Human Genetics	119	40
Archives of Surgery	661	40
Archives of Environmental Health	337	40
American Journal of Physiology	963	39
American Journal of Pathology	284	38

These were the figures reported in 1973. The percentages of rejection are on the rise. The *Journal of the American Medical Association* now claims to reject 95 percent of the articles submitted to it.

OPT TO EDIT OR FORGET IT

A more frequent occurrence today than in the past is the acceptance of an article by a journal "contingent upon editing." What do you do when you receive a letter* like the following from the editor of a journal?

> Our final concern with the paper is that the syntax and style leaves [*sic*] much to be desired. The paper as presently written is prolix, tends to be redundant between the discussion and the introduction, and is replete with jargon such as "word terminal," "different picture," "positive LE cell prep," etc. We suggest you avail yourself of someone to help you with composition before submitting the final manuscript.
>
> If you are willing to follow these suggestions, we are happy to have your final report and it will receive prompt publication.

*An actual letter, quoted with the recipient's permission.

Chances are that if you wrote and submitted the paper that way, you thought it sounded right good already. Or, possibly you knew it had problems, but you could not quite figure out what they were or how to correct them when you did spot them. Are you equipped to edit the paper now that the editor has returned it? Do you know how to spot and correct the problems? Publication of the paper depends on it—but we expect that as you complete the reading and exercises of this book you will be able to edit creditably.

Also, remember that editors and reviewers are human. Besides stacks of work, they have moods. If your writing is confusing, muddled, boring, abstruse, ad lib, et al., no matter how important the content, the editor might become annoyed and impatient and, consequently, overly critical of your manuscript; or he might ignore it altogether, except, of course, for the brief moment it takes to reject it. But, with all these human variables, the challenge of writing becomes that much more intriguing because the writing hypothesis is communication—whether or not you are able to make the external world of others' internal minds comprehend your own mind.

3. Mental Blocks, or What Stops It?

HIND-MIND HINDERED

Mention of minds brings to thought that intellectual perplexity, mental blocks (MB). One might think that one could construct something fine with mental blocks, but the truth is that the construct is blockage, stoppage, or power outage. Mental blocks to writing sometimes wear disguises. EXCUSES are available everywhere to delay writing: I must first accumulate *all* the data and all related info; why, the stuff has to be *organized* before I can put a jot on paper; Sam is gonna' call me with a piece of vital information; I am sorting my files; but the deadline is not past yet. INTERFERENCES freely occur: I do need the exercise; first, I will read the article on barnacles; if I mend the toaster tonight, I will be fresh for writing in the morning. EMPTY LIGHT BULBS form in the mind: Now where was it I saw the thing I cannot remember? what should I say next? what else is there?

An excuse is just that, an excuse; etymologically it is an attempt to free oneself from a charge. The charge to write can only be released by a charge of words from the sheer energy of the will. Inasmuch as excuses represent lack of willpower (I would rather be sailing) or motivation (who cares?), the therapy is to address oneself to that lack. Consciously review your motives for writing, whether idealistic or profane. Focus on the motive and concentrate on the reward; if necessary, frighten yourself with the consequences of the nonrewards.

If interferences are the problem, then you need to take active measures to prevent them. Time is what each of us has in the same amount each day and what each of us has too little of, whatever our purpose. Actually assigning a time for writing enables you to have others and yourself appreciate that the ritual of writing will have its appointed hour, that that ritual has importance to you, and that it should therefore have importance for, and respect from, others.

Dosage 1: Herewith, forthwith . . .

Read the proclamation to remind yourself how important messages became communicated in the early years of a great nation.

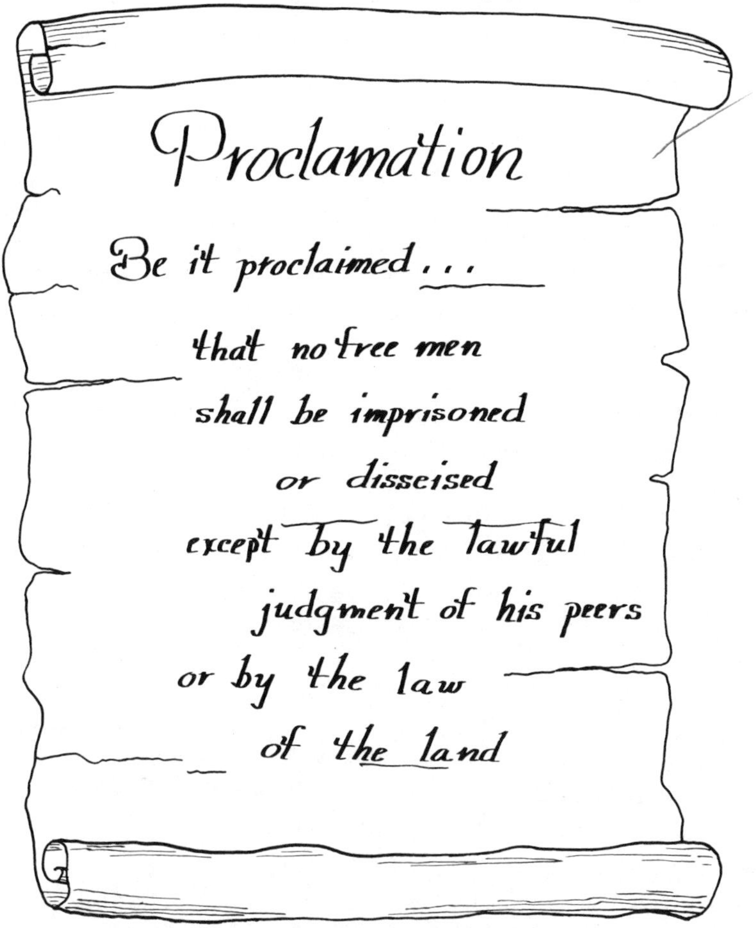

Proclamation

Be it proclaimed . . .
that no free men
shall be imprisoned
or disseised
except by the lawful
judgment of his peers
or by the law
of the land

Fig. 3-1. Proclamation (adapted from the Magna Carta).

Then write an official proclamation about your place and time reserved for writing, mentioning your desire to remain uninterrupted at said hours. Post that writ, officially, wherever appropriate–on your study or office door or the main bulletin board of the medical school. Tape a copy in your journal, on the inside cover.

If children interrupt you at home, perhaps you could have a tête-à-tête with them and lay some basic ground rules of off-limits corners and hours. Establish a favorite corner to call your own, where books can get stacked and the typewriter can be pounded to the mind's content. If the phone interrupts, ask someone else to answer it and take messages for you, or take the receiver off the hook and dial three. If the home confound cannot be remedied, then perhaps you should work in your office, taking certain precautions against interruptions by sneaking in, locking the door, using a narrow-beam desk lamp instead of the overhead lights, and putting the phone in the desk drawer. If no office is at your disposal, try the library—maybe you can arrange for a locked carrel. Other measures, more extreme, are possible when a deadline nears and a lot of product is due soon—rent a room out of town somewhere, by the sea or in the mountains. Often the new surroundings will provide new stimuli for the imagination and actually inspire your writing. Also, extreme measures usually goad you on with a certain sense of guilt if you fail to achieve what you have gone to accomplish.

Empty light bulbs are a little more complex than they appear in form. They can represent anything from a lack of self-confidence to a lack of enjoyment, and hence function as defense mechanisms—defense from feeling inadequate and defense from the pain of writing. A few practical suggestions might help you light those bulbs.

Firing the Blank

You sit looking at that first blank white page, which grows in size from $8\frac{1}{2}$-by-11 inches to a bed sheet to a drive-in movie screen to Alaska, a common symptom of "scriptophobia."

Suppose the voice of your MB taunts in your mind, "Hey, dummy, you're a scientist, not a man of letters. You made a D in Freshman English and you haven't read a novel since. How dare you think you can write an article that anyone would print?" You begin chipping at the MB by pointing out that you do in fact have letters—maybe a B.S., M.S., Ph.D., M.D., or at least your initials—and that no matter what subject matter you studied, you have been able, after all, to perform certain basic functions of the mind common to all mental disciplines, one of which is thinking. And you do think! You further point out that you are beginning a therapy in science writing called Preparation W and that if you follow its dosages step by stumble, you would be able to successfully apply your thinking ability to the English language. Finally, you could comment, "Hablo inglés," —and that is valid as a start.

If MB, not to be outdone, persists and reminds you of the rejection slips you have received from editors or publishers, point out that these are merely your membership cards in the Great Society of Rejectees. Use the rejections to advantage. Carefully reflect on each reviewer's comments on your content and style, judge whether or

Fig. 3-2. Thinker update, exposed, one hour later.

not they are valid, and then either reject them or improve your writing through their suggestions or implications.

If the treacherous MB tries the ultimate put-down–"But who cares if you write anything?"–you should review your motives until you can affirm that *you* care. True, it is easier to be enthusiastic about writing if there is an ambiance of appreciation for that activity, which of its own energy seems to create more energy and maintain the ambiance. If such an attitude does not exist around your institution, it may be because of a lack of communication. Is there some way to confer with colleagues about the research and writing you and they are doing, some way to cultivate an "atmosphere" conducive to writing, of being caught up in an energy sweep? Try starting casual interdepartmental conferences about research. Ask someone in your department or in another department to read your rough draft and give you constructive criticism on it; offer to do the same for others. Invite a small group to meet for coffee and discuss writing habits and problems. Some folks enjoy a highly structured group, as is a journal club. In such a group, someone is asked to review a given topic, to read pertinent publications, and then to report his findings to the club. Thus, one person does the research on the topic and many others benefit from that research.

When you meet a new person, ask if he is doing research; show your interest. If your institution has a staff editor, ask him to read your manuscript and to tell you the good and bad points in it. Ask the librarian to get some books on writing and to put them on a conspicuous shelf or on display. Ask the dean to arrange to have a seminar in science writing taught to the faculty.

Mind Your Cues

The preceding arrangements are psychological supports or aids to lighting the empty bulbs, to budging the mental blocks. Other practical exercises might help you get started on that Alaskan page.

First, find out when your mind is freshest. For many, the best hours are the first two after awakening. If possible, try this: (1) wake up, (2) take your mug of coffee in hand, and (3) isolate yourself for an hour or more before you even think about taking a shower or getting dressed and driving to work. Often one can accomplish more during that brief morning time than during the rest of the day. The mind works at night during sleep and is full of ideas in the morning if only there is time to trap them. Even for those who cannot coordinate their physical movements early in the morning, thoughts may come facilely. If morning is not your most productive time, then perhaps late at night, or noon, or 4:17 P.M. is. Experiment!

Dosage 2: Wheresoever, whensoever . . .

Use this form to set a time schedule and keep a record of your writing efforts. Keep the record in your journal.

Manifest Writing Schedule

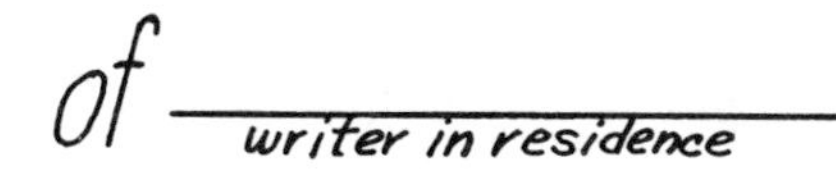

Week	Day	Minutes of Writing					Encounters
		early	midmorn	midday	evening	night	
	Sun						
	Mon						
	Tues						
○	Wednes						
	Thurs						
	Fri						
	Satur						
	Sun						
	Mon						
	Tues						
○	Wednes						
	Thurs						
	Fri						
	Satur						

Fig. 3-3. Your manifest writing schedule.

Second, if the empty bulb keeps you from shining your thoughts on paper, illuminate it immediately by writing anything. Anything. Begin with the main word of your topic, for example, "mucopolysaccharides." Then write down the very next word that comes to mind —"glycosaminoglycans." Thoughts and words are always in the mind, even when it seems blankest. Discipline yourself to put down those words whether they follow logically or not, whether they come out in sentences or not. You will find they do flow one after the other, and maybe in an order already preordained by your subconscious. Then keep writing. A later section in this book will discuss what to do with the product of this writing procedure.

Enjoyment in writing comes with facility of writing, and facility comes with familiarity and understanding of the writing tools—words and syntax. Preparation W, the elixir for writer's cramps and pains, is formulated to strengthen your mind through understanding of communication of words, and why they go together the way they go together, like two octopi, hand in hand, hand in hand, handin hand, handin inhand. . . .

TINDER BLOCKS

A preventive approach to mental blocks is to constantly store your mind with tinder blocks, stimuli and impressions that are easily fired into ideas.

In talking with friends, reading, and encountering daily stimuli, we all experience new relationships of concepts, whether by superimposing, juxtaposing, comparing, or contrasting two or more mindholdings. From such experiences come ideas, mind-play and mindreplay of how to achieve a shortcut, what to expect when two chemicals are mixed, how to account for an unexpected finding, or the most efficient way to get a specific result. Ideas are the byproduct of intellectual play; they underlie the success man has had in seeing beyond his plain/plane perspective and in probing for the answers to questionable, observable provables.

Each person must find his own optimal system for mental stimulation, whereby the combination of external events and mind-play lead to the development of new concepts, but external input is often important in precipitating new ideas. Isolation, as valuable as it is in thinking things through, may not lead to the "Aha!" that can come with bouncing thoughts off the minds of others.

Encounters with other persons can be impromptu serendipity sessions. If you are lunching with peers who understand the subject matter pertaining to an article you have in mind to write, ask them if they have recently read anything about the point you intend to make in your article. Ask them if they would have any reservations about publishing an article on such a topic. ("Now tell the truth . . . will it fly?")

If you have no specific ideas to write about, then encounters with just about anyone can help. As you meet with the family, neighbor, typist, janitor, lawyer, banker, or gas station attendant, do more than ask "Howzit goin'?" Ask about their interests and activities, listen to their conversations, see whether they have something within their kens that will overlap with your own interests. Charles Lindberg, whom you would have expected to meet among aircraft designers and in flight circles, worked for many years on the development of a design for an artificial heart. Socrates taught us to ask "Why?" Do not hesitate to ask "How?" or even "What?" in your pursuit of ideas.

Changing routines can often help generate ideas to write about. If you customarily arrive at your lab or office at 8:00 A.M. try getting there a half hour earlier and note who is there and what they are doing. You may find new vistas of friendship and ideas opening up. If you usually go to lunch at noon, try going later, and eat at different places. Take someone you hardly know with you on the adventure, and talk about such things as the need for change, new stimulation, the adversity of sensory deprivation, and the threshold of fiction into fact.

Share innovative ideas and maybe you will innovate even more. New apparatuses are always in demand. If you have animal runs, have you innovated their design? Have you adapted a piece of old equipment to a new concept? Have you tested and proved the improved efficiency of a different procedure, arrangement of equipment, or means of getting supplies (purchase, rental, "borrowing," "making do")?

Dosage 3: Gathering tinder . . .

On the schedule form from Dosage 2, plan one overt approach for encounters on each of four days of the week.

Twigs

To make the kaleidoscope of ideas within your mind become part of the experience of others, a good beginning is to write the ideas in a notebook. Set aside some time during each day in which you will spill forth from your mind those thoughts that have arisen as a result of the day's encounters and experiences.

A good method for such recording is to carry a small notebook with you to write down occurrences or ideas that transpire throughout the day. Writers often use key words to restimulate their minds about specific ideas. "Gravity control?" may mean that you question whether gravity could be used to substitute for a spring mechanism in a certain apparatus. "Morph. + mol. wt. + reactivity" may refer to a new taxonomic system you are thinking about. "Why is a tree so strong?" may lead you to thinking about a new system of construction for earthquake-resistant buildings.

In that hallowed time you set aside for writing, you can mull over those hasty notes and expand them into statements. Sometimes the

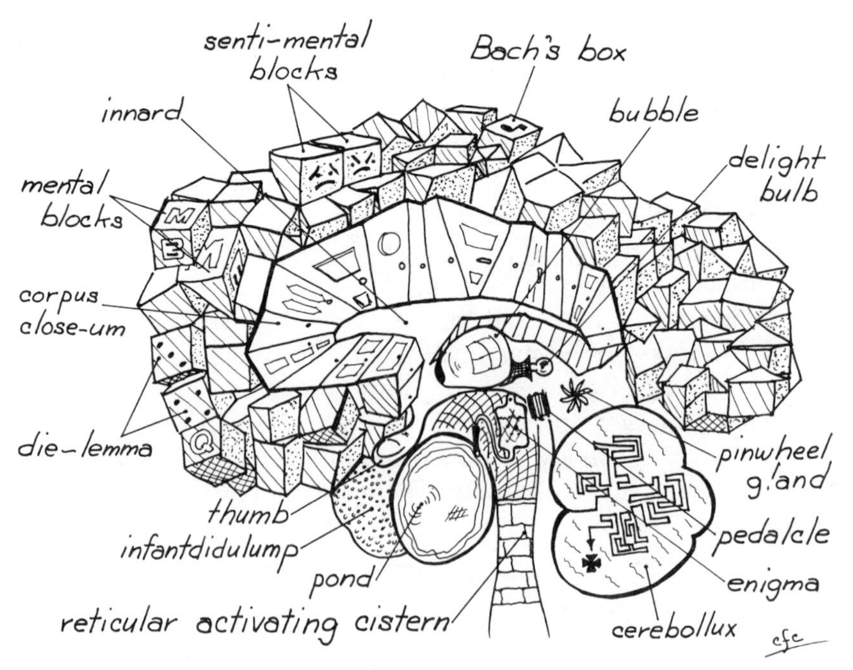

Fig. 3-4. Modern saggedtal shot of the human brain, never mind.

statements will not have much to contribute to the good of the order, but occasionally you will sense that you have come on an idea worth pursuing; at that point you might set up a file to gather further information specifically pertinent to that idea. Ultimately, of course, you may end up with a file cabinet full of materials from which to refine your ideas and produce your manuscript.

Posterity dotes on the debris of the Greats, and your notebooks and files may well be the reservoir for "support" in later life. Rediscovering notes and files that led to further notes and files might enable you to take a different turn with the ideas, to test them in light of new insights, and to even smirk a bit about early seeds and later fruits.

4. You and Babble-On

A friend, Jackie, had been intellectually desiccating in a tiny Arkansas town for two years and was desperate for a friend. Finally, a young couple moved into town. The woman would be teaching at the local school in the fall, where Jackie also would be teaching. The newcomer asked Jackie what she teaches and, when Jackie said "English," the response she got from this imagined oasis in the desert was—"English? That's easy, anybody could teach English. After all, we all *speak* it don't we?" When Jackie moments later found out that the newcomer would be teaching biology, she commented: "Biology? That's easy, anybody could teach that. After all, we all *live* it don't we?"

Well, we do all live it and we do all speak it and that, no doubt, is the problem—the reason teaching, writing about, or discussing the secrets of either is difficult. Our experience with both language and biology is too firsthand, too intimate, too subjective or personal—and that proximity keeps us from objectifying our language experience, from looking at it as distinct from ourselves. Identifying ourselves with what we say or write makes it hard to dissect our own writing because we are cutting into ourselves from the outside, having to see or hear ourselves as others hear us. But, scalpel aside for now, we will consider a few theoretical concepts.

Language and writing are part of the great social pastime, which is not sex or sports or even drinking—but communicating. Communication seems to exist for most levels of being. Plants, we now are assured, communicate with themselves via some mysterious medium within the given environment, presumably involving chemical, mechanical, or electrical impulses.

And thanks to Leeuwenhoek, who developed the microscope, we have known since 1676 how to receive information from microorganisms, although, despite the length of the interim, we are still inept at communicating *to* those organisms. Man is such a solitary creature in his proscriptive universe that his means of communication to the wee-be's is the medium of dysequilibrium-threat to their environment. Consider, for example, a product given the name Belchaseltzer. Bacteria that live in the gut receive our message through the Belchaseltzer and have no option for response but to squelch the belch. Communication in this case is the mere transmission of information, the sending out of a message. However, communication can also be the more complex cycle of transmission-message-reception-recognition-response. This process leads us up the ladder to canines and man.

Information, the stuff of communication that is explicit and implicit, includes sense perceptions, symbols, signs, abstract ideas, scientific data, diagnoses, and Kathy's home phone number. The mystery of communication is that we each can act on information in our idiosyncratic ways. The newcomer in Jackie's story, for example, used the explicit datum that Jackie teaches English to implicitly communicate the information that she thereby considers Jackie untalented and intellectually inferior. In turn, Jackie's implicit message to the newcomer was, "You're stupid." Hence, we can be conditioned or easily provoked to react with emotional override, to ignore the explicit information and respond to the implicit information.

How do we know how our messages will be received? We will always have doubts as to whether or not our messages are interpreted as we conceive of them. Our sense of communicating begins in utero, is trip-hammered by the first outslapped cry, and becomes transformed into persona with age. Man communicates with his fellow person by letter, by electronic media, and by the social cues of the blush, smile, wink, bowed head, and, alas, flatus begattus. He receives a response or perceives fresh communications through the same cues.

In the givens of his milieu, by eye, ear, nose, and throat, and by cute systems many are unaware of, including thought transmission, proprioceptive cueing, and dyspepsia, the crying-out-loud infant learns that putting out the wail provides surcease of, or release from, the agonies of his vale of tears. Later in life he learns that the degree of his signal should be modulated to correspond implicitly to the degree of his distress. For a pin sticking in his derriere, he really lets go; but, for a wrinkle in the sheet, a whimper suffices.

KIND OF GRAPHICALLY SPEAKING

In his book on body language, *Kinesics and Contexts*, Ray Birdwhistell presents a notational system to represent facial and body expressions that add meaning to a word or statement. This kind of code is typical in cartoons.

Kinegraphs*

Blank faced

Wink

Shifty eyes

Nose wrinkle

Right sneer

Ear wiggle

Fig. 4-1. Facial expressions are signs of communication.

*Ray L. Birdwhistell, *Kinesics and Contexts* (Philadelphia: University of Pennsylvania Press), 1970, p. 260. The kinegraphs illustrated are only a few in Dr. Birdwhistell's notational system.

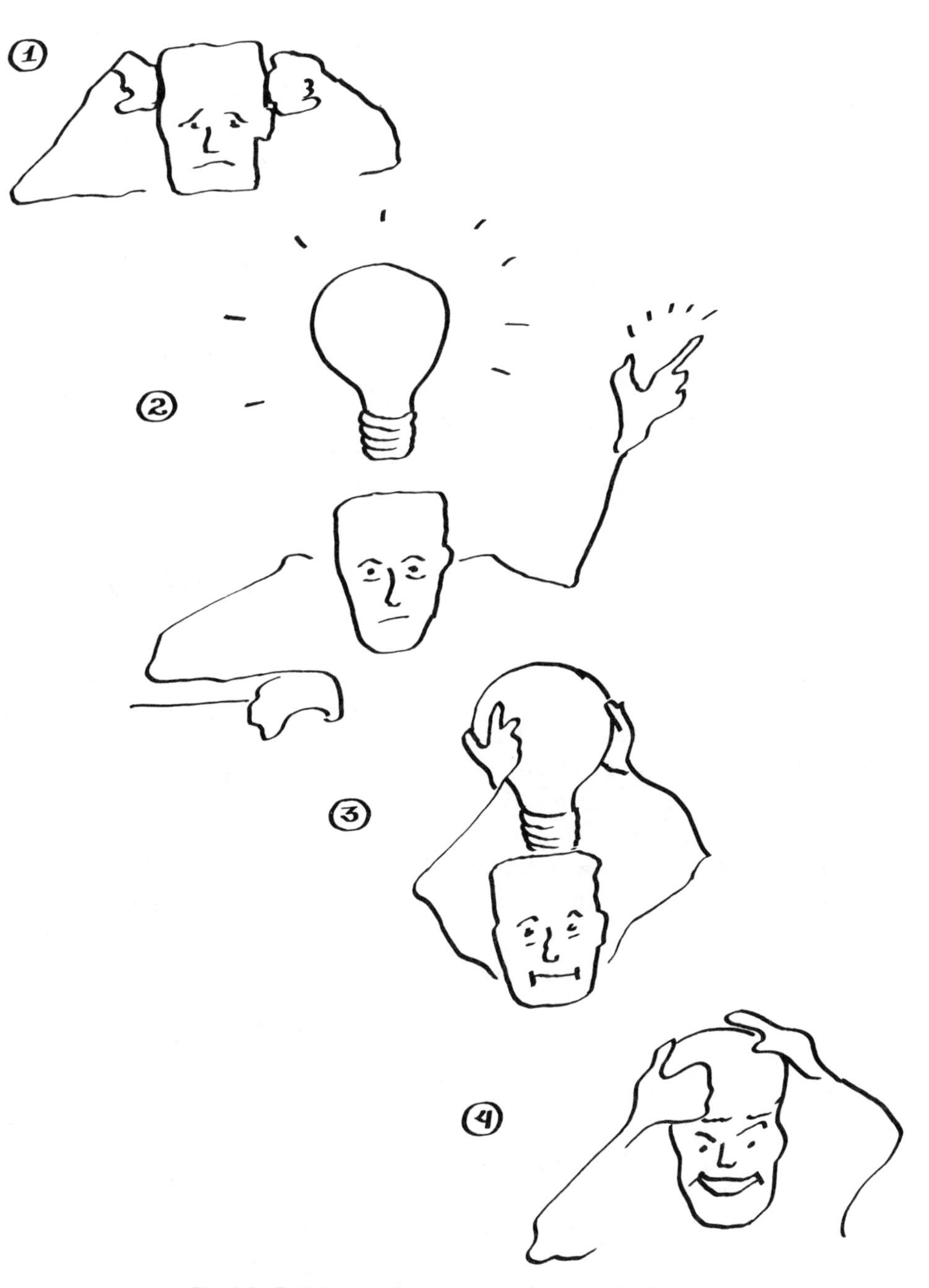

Fig. 4-2. Facial expressions are signs of communication.

Dosage 4: As I said . . .

Enter this Dosage in your journal. Do not read on until you have completed the Dosage.

To advance your own awareness of the relationship between gesture and language, try the following exercise. Write down a conversation you recall or make up a dialogue. For example, write a dialogue in which a physician tells his patient his diagnosis and discusses its implications for the future. Then, make up your own set of kinegraphs and draw them in where appropriate in the dialogue. There will probably be at least one kinegraph per sentence.

When you have finished this Dosage, you should be more aware of the role that nonverbal (or phaetic) communication plays in spoken language. Do not continue reading until you have taken the full Dosage.

A set of kinegraphs and narrative written by an anatomist[†] are presented below as an illustration of how someone else perceives phaetic communications. The patient is very alert to the "signs" of meaning because he is overly anxious about the forthcoming message. The preoccupied physician seems oblivious to such signs and perceives neither the patient's distress nor the miscues he is giving the patient.

Kinegraphs

⌣⌣ Brow moderately furrowed (concern, thoughtful)

^ ^ Brow intensely furrowed (extreme concern, worry)

<—> Tight lips (tense, restraint)

⌶⋞ Uneven eye squint (pain)

†Printed with permission of Robert Dyer, Ph.D.

Body stretch

Medial brow contraction (tense, concern)

Finger motion (giving directions)

Sitting in a rigid posture (ill at ease, tense)

Flashing eyes (urgency)

Eyes wide open (intent listening or staring)

Nod of head (affirmative)

Relaxed face (calm, content)

Eyes focused (attention being given)

Brighter face (delight)

Lively face (optimism)

Confused look (lack of comprehension)

Jovial face (relief)

Wave of the hand (good-bye)

Straighten up (repossession of self)

Handshake (interview ended, friendship)

Quick departure (relief, joy, lighthearted)

Narrative

I had just received a very thorough physical examination and while slowly getting dressed, I contemplated the upcoming confrontation with my physician. I had requested a physical because of a general overall poor feeling and occasional sharp obnoxious pains in my lower right side. As I bent over to tie my shoe, I felt an excruciating pain in my left side. I quickly straightened up and placed my hands in the small of my back and stretched to relieve the pain. I then went to the office to await my diagnosis and prognosis.

As I entered the office, the doctor was engaged in a tense conversation on the telephone. He motioned for me to be seated, and continued his conversation. His tone of voice was serious and I sat on the edge of my chair in a rigid posture as I overheard him discussing an upcoming surgical procedure. He was insistent upon an immediate operation, since the life of the patient was in danger. I became very concerned that he was discussing my case, and listened as intently as I could.

With a nod of his head he placed the receiver on its cradle and turned to face me. He had a worried look on his face and I had the feeling he was staring right through me. The expression on his face relaxed and his eyes focused upon mine. I was prepared for the worst.

"Oh yes," he exclaimed, his face becoming brighter and more lively, , "we need to say a few words about your condition!"

I was confused by his apparent lack of concern over my fate.

He continued, "I have carefully evaluated your situation and can honestly say that you have nothing to worry about." He was almost jovial as he pronounced my condition and stated, "I think a good dose of milk of magnesia will cure all your ailments."

Needless to say, a sense of relief swept across me and I felt as though I had been given a new lease on life.

"Now you can go home and allow me to finish a problem at hand. I've got an emergency appendectomy to perform."

I quickly straightened up, shook hands, and quickly left the office and my cares behind.

5. You, the Writer

If in speaking we depend not only on words but also on gestures and feeling tones, in writing we have only words and they must somehow supply the gestures and feeling tones. As difficult as communication is, you are at an even further disadvantage with the written word because your listener is not standing there before you to give you an immediate response to your message. He is maybe a thousand miles away, a whole culture away; and, as he reads your article in a journal, he has no way of transmitting gestures or glances to you to let you know how he is reacting to your message, whether favorably or unfavorably, or whether or not he even understands it at all. Writer and reader are analogous to the scientist and the microorganism under the microscope—the reader being in a position only to receive the message. You as writer are in the position of the Belchaseltzer: you can send the message and only hope the reader squelches his belch.

That notwithstanding, all is not hopeless and all need not be inaccurate or doubtful communication. Both explicit and implicit information can, in fact, be transmitted. To transmit information successfully, you must understand the signs and symbols that are common to most people and understand the patterns of putting them together that are comprehensible to most people.

GAP GAPING

Do most people comprehend the same words in the same way, and do most people put them together in the same way? Yes and no.* Most people use the same basic patterns of language, the way words follow one another in a sentence; these patterns are called syntax. But, for any series of combined words, the idea comprehended can

*Our remarks are restricted to the use of American English. We are not making intercultural generalizations here. The patterns in French, for example, are different from those in English or Spanish.

Fig. 5-1. Lapsus demonstrates the squiggle line, originated by one of his forebears. Copy editors use squiggle underlines to indicate that words should be set boldface.

vary widely among folks because, for any given word, each person has an entirely different set of learning experiences and associations through which he came to know that word. Explore that idea by taking the following Dosage.

Dosage 5: Honestly speaking . . .

Think about the word "honest." Experiment with the word among several friends.

(1) First, ask each friend for a simple definition of "honest." How many different definitions do you get?
(2) Next, ask for a description of a situation in which a man demonstrates by his actions that he is honest. Do you get the same action in all responses?
(3) Describe a situation that you perceive as borderline honesty and ask your friends (without stating your own opinion) whether or not the person was honest in his action (for example, a "white lie," hedging on income tax).

We recommend "polling" several persons separately and two or three others in assembly. Record their responses in your journal. You might want to try other concepts, such as loyalty, fidelity, hot, aqua, intelligence, electricity, hypertension, protein, sexy, nutrition, or authorship of an article.

We all use words and assume our audience understands them in just the way we understand them; yet, that would be to assume that we have all shared identical life experiences in toto, that we have all shared the same judgments of "honesty." Obviously, we have not, and the differences produce a "communication gap." In speaking, we can mediate the comprehension of our words with gestures or voice tones, but we cannot do so in writing. Hence, in writing, and especially in science writing, we must severely discipline our choice of words, selecting only the most precise words.

To illustrate characteristics of the communication gap, two Dosages are offered below. The first requires only your imagination, the second the cooperation of a friend.

Dosage 6: Subjective objectification . . .

As an exercise of awareness in communication, try to objectify language. Part of objectifying language is getting it outside of yourself and of your own meanings, which can be aided by seeing yourself address yourself to your reader. So first, take in hand a paragraph you have written, perhaps from a memo or a draft of an article. Read it through and ask yourself: What did I want to say in this paragraph? Do not change any of the words. Then, imagine a person who might be a reader of that paragraph right in front of you, his back to you, the paragraph in his hand, ready to read and perceive your message. Now, imagine that the reader slowly and reflectively reads each word of your paragraph.

Remembering what you learned through Dosage 5 about how different people have different nuances of understanding of words, continue with the scene and listen to your words as your reader reads them. As you watch, can you tell if the reader understands every word as you meant it? What interpretations might the reader make that you did not intend? Do gaps in the meaning of words become apparent now?

Remember, you are outside of the reader's consciousness and you cannot communicate with him. You cannot tap the reader on the shoulder to tell him that he is not getting the message, nor can you tell him if words were left out that would clarify the meaning, because the paragraph is in an article in a journal in the listener's hands, a thousand miles away.

Record the potential communication gaps in your journal.

To make yourself more sensitive to communication gaps in daily activities, practice this objectification of language with others when you are not involved in the conversation, and record your reflections in your journal. At a meeting, for example, briefly remove yourself from the dialogue and watch the interactions. Is the group understanding the dean's message? If two members are disputing a point, are they really interpreting one another correctly to begin with? Are they communicating explicit or implicit information? In a classroom situation, is the teacher expressing himself precisely? Does he really understand the student's objection or question? When a new employee is introduced to his work in the laboratory, have his instructions been explained precisely and has he understood those instructions in relation to all the new information in his new environment? Has the department chairman made a new faculty member's teaching, clinical, and research obligations explicitly clear or are there implicit assumptions that could lead to serious communication gaps?

This approach is not meant to make you critical of others, but to ease you into objectifying your own language habits. Our tendency is to note the weaknesses in the communications of others, while overlooking those weaknesses in ourselves. Use the approach on yourself while interacting with others and be strict with yourself in assessing your own communication gaps. Toward whatever end one chooses to write, knowledge of one's intended and implied meanings is a strong advantage and should be a personal imperative. Unless we can perceive ourselves as others do, we can never know exactly what we are communicating to others or if, indeed, we are communicating at all. "Communication" etymologically develops from the concepts of "union with" (com-union), or "common," the "sharing of." The sharing in common of an idea through a means that is already in common (language) is the essence of communication.

Dosage 7: Art of wrote . . .

Now try the following game, which should help you analyze some of your own language and thought habits. Do *not* look ahead to the next page or step. It is important that you complete each step of the exercise before reading further. *Complete* each step *before* reading the direction given in the *successive* steps. Have a friend or foe nearby who is willing to cooperate with you for about ten minutes.

Step 1: On a blank page in your journal, preferably on the left-hand side, make a drawing, any kind of drawing–abstract or representational. Limit yourself to three minutes. Do not read the next direction until you have finished the drawing. To encourage your honest adherence to our request, we have hidden Step 2 in the Glossary of this book, alphabetically inserted under "S."

Step 2: (See Glossary, p. 279.)

Step 3: (Do *not* read this step until you have completed Steps 1 and 2.) Now, get the friend or foe to participate. Do *not* let the participant see your drawing. Ask him to execute your directions on a sheet of paper or on a blackboard. Read each step of the directions to him, one by one. Have him complete each step before you read the next direction. Do *not* fill in any details, nary a word, that are not written down. You may reread a direction but the participant may not go back to a previous step and change the drawing after he has proceeded to another step. Do *not* respond in any way to any of his drawings; maintain a poker face. (If you cannot trust yourself, your eyes, or your occasional nervous twitch, turn your back so you cannot see the drawing the participant is producing.)

Step 4: When your cooperator has completed all steps of the directions, reveal the drawing in your journal and compare it with his. Are they identical? If not, why not? Where were the directions incomplete or misleading?

Perhaps your directions in Dosage 7 were so clear that your drawing was reproduced exactly; however, for most this would not be the case. The game was meant to illustrate that there is usually a gap between what is in your mind and what is in the mind of your reader whenever you try to communicate a thought. Sometimes this gap is the fault of the speaker, sometimes of the audience. You may be absolutely certain in your mind that you have completely and correctly relayed all the information necessary for full comprehension of what you are saying, but with all the contingencies involved– for example, your reader's past experience with a word, his knowledge of grammar or of the subject matter, his gastritis, your regional language quirks–you cannot be sure of his interpretation. So, without belaboring the details of your subject matter, you must figure out how to relay all the essentials of your information for communication.

Dosage 8: That is not what I meant . . .

Below is an example of a drawing done by a physician–one gone askew (the reproduced drawing, that is)–and an analysis of why. (Have you reviewed your Materials and Methods Section lately?) You, now, are to assume the participant's role. Therefore, do not look ahead to see the original drawing until you have completed the physician's prescription for reconstructing his drawing. Draw the figure in your journal.

The Directions

Construct a figure, with the use of a pen, composed of the following:

Step 1: Draw a line, all points of which are equidistant from a central point.
Step 2: Draw a second line, the length of which is equal to one-fifth that of the first line; the nadir of this line must be located a distance of one-half the radius of the line in Step 1 from any point on the line described in Step 1.
Step 3: With a pencil, draw a straight line connecting two points on line 1, the center of which is at a distance equal to an imaginary line drawn from the center of line 1 and any point on line 1.
Step 4: Now draw two arcs, the apices of which are directed toward the nadir of line 2, and whose limits intersect, respectively, points one-eighth and three-eighths the distance from each end of line 3.
Step 5: Then erase the pencil line (line 3).
Step 6: At a point on line 1 which is slightly to the left of a point on line 1 which is at the extreme distance from the nadir of line 2, draw a curlicue.

Toward a Definition of Skewness

Depicted here is a drawing a participant devised from the directions in Dosage 8. Does it resemble your drawing? It does not resemble the original drawing. The original drawing can be found between "feel" and "female" in the Glossary of this book.

Fig. 5-2. A participant's product of the directions in Dosage 8.

The drawing started off on the wrong foot in Step 1: "A" central point? That means any central point, and in this case must mean the center of the page. How long is a line? Oh–a *continuous* line?

Well, even if a circle had been drawn and, based on Step 2, a curved line had been drawn in the middle of the circle, would not Step 3 have tripped you up?

In Step 6, how can a point be at "the" extreme distance from the nadir if the nadir is in the center? In which direction should the curly cue?

6. You, Reader

YOU READ IT

Dosage 7, the art-of-wrote exercise, should have revealed something about another's ability to comprehend what you write. How well do you read what others have written?

Dosage 9: T, F, or ?

Read the paragraph below. Then answer each question successively. Do not return to a question once you have answered it. If the statement is true, circle T; if the statement is false, circle F; if any part of the statement is unknown, that is, information not supplied by the paragraph is needed to ascertain truth or falsehood, circle the ?.

The head of the Fulton Medical School entered the pathology laboratory of one of the faculty, where he found three technicians playing cards. One of them was Fred Young, brother-in-law of faculty member Dr. Henry Watson. Dr. Watson, incidentally, often worked late. School policy did not specifically forbid gambling on the premises, but the dean had expressed himself forcefully on the subject.

Questions

(1)	In brief the story is about the head of a medical school who found three men playing cards.	T F ?
(2)	The dean walked into the laboratory of one of his faculty.	T F ?
(3)	School policy forbade playing cards on the premises after hours.	T F ?
(4)	Although the card playing took place in a faculty member's laboratory, the story does not state whether it was Watson's.	T F ?

(5) Gambling on the premises of Fulton Medical School was not punished. T F ?
(6) Watson never worked late. T F ?
(7) Fred Young was not playing cards when the dean walked in. T F ?
(8) School policy forbade gambling on the premises. T F ?
(9) Three technicians were gambling in a faculty member's laboratory. T F ?
(10) Although the card players were surprised when the head of the school walked in, it is not clear whether or not they will be reprimanded. T F ?
(11) Dr. Henry Watson is Fred Young's brother-in-law. T F ?
(12) The dean is opposed to gambling on the premises. T F ?
(13) Fred Young did not take part in the card game in Dr. Watson's laboratory. T F ?
(14) A medical center's chief executive found three technicians playing cards. T F ?

(See the Appendix for answers.)

In case you think you were tricked, you were not. The fact is, we all read with a great deal of inference, often excusing our inferring as "reading between the lines" and developing a habit of in(ter)-ference of the message. To infer that an author, specifically a science writer, means more than he has said is a potentially dangerous assumption, especially when the interpreted message is applied clinically. For an author to imply more than his words say is equally risky in communications. Sometimes, unbeknownst to the author, his words imply more than he intends, hence the urgency of critical self-appraisal in writing.

Another Dosage against inference is supplied below. More alert now to the vacancies of words and to your eagerness to help the author out, see how you score. Follow the same directions as you did for Dosage 9.

Dosage 10: Try again . . .

Dr. William Phillips, the research director of a midwestern university, ordered a crash program for development on new research. He gave three department heads authority to spend up to $50,000 each without consulting him. He sent one of his best department heads, Professor Harris, to NIH in Bethesda with orders to learn the newest research technology. Within one week Harris produced a highly promising approach to new research at the university.

Questions

(1)	Phillips sent one of his best department heads to NIH.	T F ?
(2)	Phillips overestimated Harris' competence.	T F ?
(3)	Harris failed to produce anything new.	T F ?
(4)	Harris lacked authority to spend money without consulting Phillips.	T F ?
(5)	Only three department heads had authority to spend money without consulting Phillips.	T F ?
(6)	The research director sent one of the university's best department heads to NIH.	T F ?
(7)	Three men were given authority to spend up to $50,000 each without consulting Phillips.	T F ?
(8)	Only four people are referred to in the story.	T F ?
(9)	Phillips was research director of a university.	T F ?
(10)	Whereas Phillips gave authority to three of the best department heads to spend up to $50,000 each, the story does not make clear whether Harris was one of these.	T F ?

YOU, READER

Were Alice here today, she would be told by the Rabbit that pencils are obviously made in Pencilvania. The measure of literateness in that happy land is manifest most in reading; in fact, Rabbit and Alice would both be quick to point out that reading is critical to the success of pencil use, vanity images in mirrors, and reflections on the general state of being. Through reading one develops awareness of present states, changes, projections, predictions, and the seminal explanation for one's remarkable loss on the stock market. Reading provides the underpinnings for rationale; it serves as a guide to information needed in normal pursuits of daily life; and within it and through it the reader may find his emotions put upon, his anger piqued, and his pleasure-cup filled.

But, *the* reason reading is essential to the writer is that it is his means of developing awareness of language itself. You may object: "I read, I read, I read the journals, and they tell me what I want to know." They tell you part of what you want to know; they give you explicit scientific information, but not necessarily in an explicit manner. If you have had to reread an article or paragraph and concentrate too hard to get the information, perhaps you have assumed that the idea was profound or complex. But the fact is that no matter how compound the idea, it should not confound the reader if it is well written. The best science writing should read quickly, energetically, in straightforward language. These qualities are not descriptive of typical science writing, if we are honest in our appraisal.

What happens with the writer as reader is that he begins, however subconsciously, to imitate those language patterns he encounters. A Georgian speaks like another Georgian, a Yankee like a Yankee, a Texan like a Texan, and an Englishman like a pub-bub. The scientist-writer tends to write in the style of all the other science articles he has read, and in this way habits of language develop but do not necessarily evolve with the ever-spiraling consciousness. A casual misunderstanding of the medium can propagate convention, good or bad, such as the misuse of "disinterested" to mean "uninterested"; or of "disorganized" to mean "unorganized"; or of the word "column" by the director of a research program to mean "two lines at the bottom of a newspaper page," which resulted in a twelve-inch, weekly newspaper column agreed on by his science writer and the editor of the newspaper.

To unlearn mistakes and bad habits, the writer can begin by reading good writing, which means going outside the area of science writing to other materials, whether factual or fictional. A brief list of books on a variety of topics and interests is given below. If some of them appeal to you by subject matter, we recommend that you read them for examples of style. You might also take a subscription to a lay magazine (other than the weekly news magazines) with well-written articles that are well-developed essays in intelligent and straightforward language. A few suggestions are *Harper's*, *Atlantic Monthly*, and *The New Republic*.

Recommended Books

Albert Camus, *The Stranger* (twentieth century existentialist novel)
G. Flaubert, *Madame Bovary* (nineteenth century French novel)
Ernest Hemingway, *Short Stories* (twentieth century)
Ken Kesey, *One Flew Over the Cuckoo's Nest* (twentieth century American novel)
Flannery O'Connor, *Everything That Rises Must Converge* (twentieth century American, short stories)
Dusto Popov, *Spy Counterspy*, 1974 (autobiography)
Carl Sagan, *The Dragons of Eden: Speculations on the Evolution of Human Intelligence*, 1977
Voltaire, *Candide* (eighteenth century French satire, adventure; translation by Blair, Bantam paperback)

THE READ WRITER: ACQUIRING A PERSPECTIVE

In many African cultures, one customarily communicates a gripe or settles a dispute through the use of proverbs and parables. The proverb embodies an accusation and a lesson but depersonalizes the dispute by referring to animals or to a legendary person. Because both parties can talk "about" an issue, the accused person can follow the logic and the justice described in the parable and then apply the justice to himself without the anger and resentment a direct accusation might have evoked. In John Gay's narrative about life in a Liberian village, *Red Dust on the Green Leaves*, as Flumo's sons

> . . . listened to their elders, they saw that language is power. Their father was often asked by townspeople for his advice, which he rarely gave directly, in clear, outspoken language. He spoke in proverbs to those who came to him for help. To a person who

was reaching beyond his capacity, he said, "A short man does not measure himself in deep water." To the woman who was vexed with her husband for not making a bigger farm, he said, "Can rabbit do the work of elephant?"*

The writer will need the same sort of distance from his own words to be able to evaluate his writing objectively or to receive constructive criticism without hurt feelings or wounded ego. Dosages 5–10 were steps toward that objectivity.

Once words are written, they are, in fact, distinct from the author and do take on their own existence and meaning. Therefore, if a sentence does not adequately express a writer's idea, his confidence should not be destroyed and he should not consider himself unintelligent or inadequate. The sentence is not necessarily a reflection of his idea, but rather a misreflection of it, which can be improved. However, human nature being abrupt as it is, others are likely to consider a person's writing faults as symptomatic of a lack of ability or intelligence in areas other than writing. Since human nature is not easily improved but communication habits are, you might bear in mind that "It's easier to trick a new dog with old treats than to crave a drivel through the need of an isle."

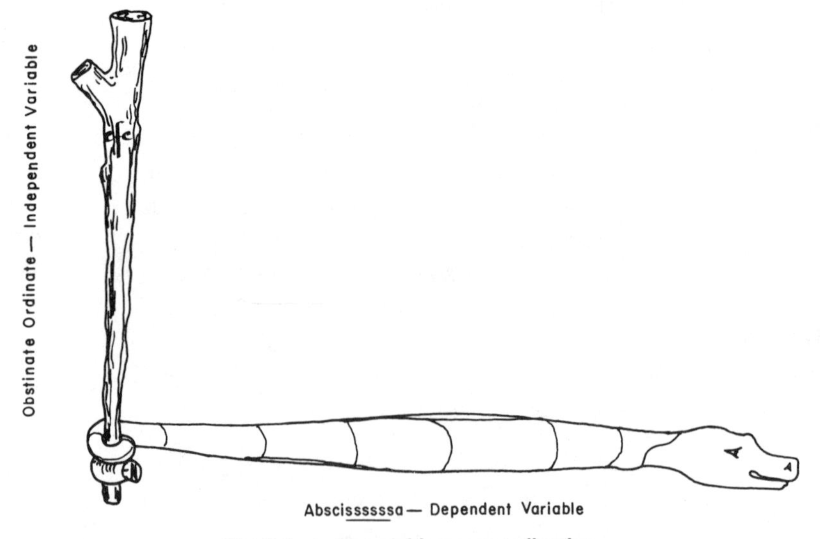

Fig. 6-1. Lapsus and Logos coordinating.

*John Gay, *Red Dust on the Green Leaves.* Thompson, Connecticut: Interculture Associates, 1973, pp. 66–67.

PART II.
WORDS AND COMMUNICATION: HEAR-SAY, SAY-WRITE, READ RIGHT

7. You Write With What?

A CHISELING DISCOVERY

In explanation of how he came to make the statue of "Night," Michaelangelo replied:

> "I had a block of marble in which was concealed that statue which you see there—the only effort involved was to take away the tiny pieces which surrounded it and prevented it from being seen. Every piece of stone or marble, whether large or small, has a statue or effigy within it—but of course one must know exactly how to carve away only that which hides the statue, and this is very dangerous in that one may take away too much or too little. For anyone who knows how to do this, nothing could be easier."*

Michaelangelo's description of his act as sculptor is a good analogy for most creative work, and because these words are expressed in the way that they are, the hearer of them perceives the material and the act of sculpting in a special way. Michaelangelo's phrasing makes one view the stone as though it, solid material, were alive with an energy of being and the sculptor only cooperates with nature. To consider the statue as already in the stone renders: (1) the stone a protective concealor of an entity waiting to be discovered, (2) the statue alive and fully formed, with an autonomy of being, and (3) the sculptor both a creator and a midwife to nature because his insight conceives the statue which he makes visible to man literally by doing the chiseling.

In a science laboratory, the researcher chisels his statue out of the biochemistry of the tissue. He chips away at his data, others' data, and new data—until he sees the form inherent there, inviting him to reveal it. A striking example of this phenomenon is described in *The Double Helix:*

*Quoted in: Giovanni Papini, *Michaelangelo: His Life and His Era*, translated by Loretta Murnane. 1952, p. 275. © 1952, E. P. Dutton. By permission of E. P. Dutton, New York.

> After lunch I was not anxious to return to work, for I was afraid that in trying to fit the keto forms into some new scheme I would run into a stone wall and have to face the fact that no regular hydrogen-bonding scheme was compatible with X-ray evidence. As long as I remained outside gazing at the crocuses, hope could be maintained that some pretty base arrangement would fall out. Fortunately, when we walked upstairs, I found that I had an excuse to put off the crucial model-building step for at least several more hours. The metal purine and pyrimidine models, needed for systematically checking all the conceivable hydrogen-bonding possibilities, had not been finished on time. At least two more days were needed before they would be in our hands. This was much too long even for me to remain in limbo, so I spent the rest of the afternoon cutting accurate representations of the bases out of stiff cardboard. But by the time they were ready I realized that the answer must be put off till the next day . . . .
>
> When I got to our still empty office the following morning, I quickly cleared away the papers from my desk top so that I would have a large, flat surface on which to form pairs of bases held together by hydrogen bonds. Though I initially went back to my like-with-like prejudices, I saw all too well that they led nowhere. When Jerry came in I looked up, saw that it was not Francis, and began shifting the bases in and out of various other pairing possibilities. Suddenly I became aware that an adenine-thymine pair held together by two hydrogen bonds was identical in shape to a guanine-cytosine pair held together by at least two hydrogen bonds. All the hydrogen bonds seemed to form naturally; no fudging was required to make the two types of base pairs identical in shape.*

This phenomenon of insight is also experienced by the clinician who studies the signs, symptoms, and laboratory reports before diagnosing a disease.

Likewise, in writing, the stone is there—the words—with an image or meaning waiting to be chiseled out by an author's arrangement. As the writer chips away the excess verbiage, the statue or idea is dis-

*From *The Double Helix* by James D. Watson. © 1968 by James D. Watson. Reprinted by permission of Atheneum Publishers.

covered in a unique arrangement of words. Even editing can be a type of sculpturing inasmuch as a precise message is concealed within the words, a seemingly muddled and obscure block of meaning in a manuscript and in an author's mind that can be rendered into a living idea. The editor works with the author to refine the message out of the muddlestone.

PENNING THE TOOL OF THE REALTY

To chisel out good writing, you have to know your material and tools. You have a favorite pen, you say? Well, 'tis claimed that the pen is mightier than the sword, but we were thinking about words as tools. In writing, words are the basic elements of communication; we can pile them up and arrange them in many ways to form compounds and confounds of thought. Awareness of words, consciousness of words–of what words are and how they function–is critical to good writing, and this awareness usually results in such an enjoyment of words for the author that writing can actually be fun. Word-awareness includes word-origin, word-relatedness, word-representativeness, and word-suggestivism.

What is a word? Merely giving a definition of a word does not bring a person any closer to understanding *what* words are. Words are a part of us by cultural heritage; we cannot blot them out of our minds. Thus, to understand words is to understand an intimate part of ourselves. Such comprehension requires serious reflection about the concept; perhaps one can best approach words through oneself.

For Dosage 11, try to respond to the five probes given below and write your thoughts in your journal. If the probes perplex you at first, think about each one slowly and carefully. After you have written some thought about each probe, continue reading the text.

Dosage 11: A word is . . .

Explain (in your journal):

Probe 1: What words are (a) in general, to most people; (b) specifically, to yourself.
Probe 2: How you relate to words. (What do you *do* with

Fig. 7-1. In a pig's eye, the pen *is* mightier.

them? Are you comfortable with them? Do you control them or do they control you? Do they define your internal reality? Do they limit it? Do they give you a particular understanding of yourself or of the world?)

Probe 3: How words relate you to external reality (the world out there).

Probe 4: How your internal reality described in Probe 2 relates you to the external reality described in Probe 3.

Probe 5: How to *communicate* your internal reality described in Probe 2 to the external reality described in Probe 3.

Thinking of the above as analogous to a laboratory experiment might help clarify the concepts. In an experiment there are unknowns—both internally (in the scientist's mind) and externally (in the experimental materials)—and so the scientist posits a hypothesis about his materials. The hypothesis is a tenuous bridge between the internal and external realities. To pursue the assumption about the experimental materials, the researcher develops a method for testing and measuring the conditions of the materials. Are words some sort of tenuous bridge or hypothesis between realities?

Words are first, most simply, and in general, *sounds*. In radio, variations in microamperes in the electromagnetic spectrum are put forth from transmitters and captured and channeled by circuitries, resulting at the turn of a switch in the output of sound, including music, speech, and, not necessarily exclusive from the aforementioned, noise. The sound produces within us an effect. With speech-sound, the effect is the formation of images produced in our minds by our interpretation of the changes in tone, pitch, and force.

The sound is therefore a *symbol*, a representation of something beyond itself. Experience is the great conditioner of how one interprets the sounds within the mind. Thus the circuitry is, simplistically, that one person, having a given series of images he wants another to have, converts those images to transmittable sounds. As simple as that system would be at the mechanistic level, it is richly complex at the physiological and psychological levels.

The sound symbols of speech can also be perceived through visually observable symbols. The communicator arrives at the need to represent the sounds themselves, and the alphabet enters. Enter next the use of the alphabet to record the sound-for-images flow as it outpours from the mind or springboards from the tongue.

In the poem "A Noiseless Patient Spider," Walt Whitman describes the mind's process of attempting to communicate with external reality.

A Noiseless Patient Spider*

A noiseless patient spider,
I mark'd where on a little promontory it stood isolated,
Mark'd how to explore the vacant vast surrounding,
It launch'd forth filament, filament, filament, out of itself,
Ever unreeling them, ever tirelessly speeding them.

And you O my soul where you stand,
Surrounded, detached, in measureless oceans of space,
Ceaselessly musing, venturing, throwing, seeking the spheres to
connect them,
Till the bridge you will need be form'd, till the ductile anchor hold,
Till the gossamer thread you fling catch somewhere, O my soul.

Like the spider, the poet is isolated in the measureless internal reality of the spirit, "musing" or thinking, seeking the symbols (words) that will bridge his thoughts solidly, securely ("till the ductile anchor hold"). The gossamer thread is the word, airy sound that it is, and it resembles the filament, filament, filament of the spider as the visual symbols flow from the author's pen.

WEBBED IN THE UNCONSCIOUS

Remarkable about the creative process of writing is that so much of the formative activity seems to occur in the unconscious. Whereas technique can give us control, conscious control, over the written words, the insight to ideas often seems to be triggered in the unconscious.

*Walt Whitman, *Leaves of Grass and Selected Prose*, edited by Sculley Bradley. New York: Holt, Rinehart and Winston, 1949, p. 370.

Be alert to anything that your mind is willing and able to give. Expand the horizon of your conscious thought beyond points you formerly considered the absolute limits. And as you cast your line far into the outreaches, you will find you can cast your lines down on paper with remarkable ease, freshness, and perhaps even cleverness.

Dosage 12: Free word-association

On a blank page in your journal, write the sound-image (word) that represents your favorite color. Next, timing yourself for three minutes, write down *any* and *all* words that come to your mind. The words need not have any apparent relationship to the color, and you should not consciously control their order of occurrence or put them into sentences. Write the words in vertical columns. When you have finished with a color, write down the word "protein" and free associate for another three minutes.

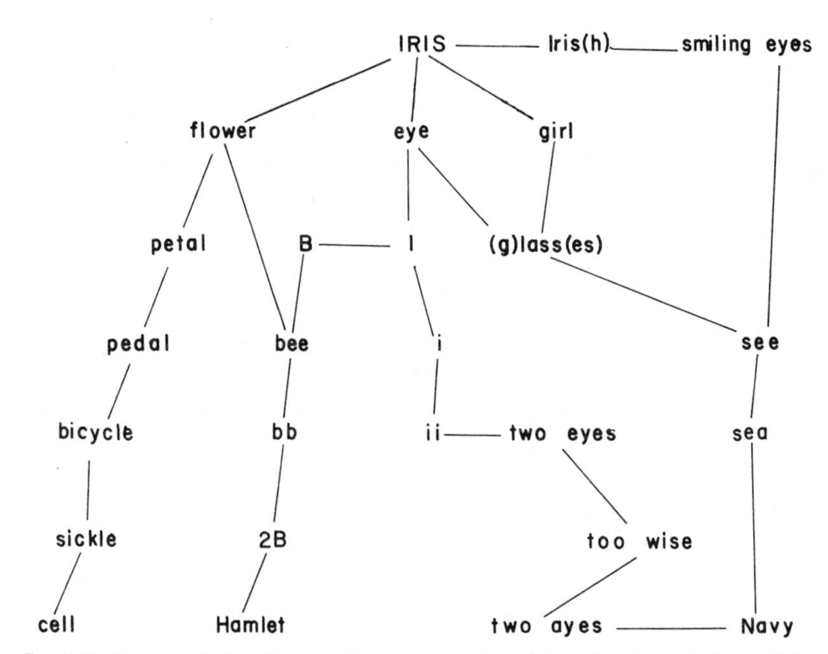

Fig. 7-2. Free-association diagram, showing extension of thoughts from single word, ***iris*****.**

Practice this free association process with individual words as often as is necessary to get the sound-for-images flow easily bridging your mind to paper. Whenever you come to a mental block in writing, especially to scriptophobia, try putting down the key word of your subject matter and do a free association exercise. New relationships in your subject might come to mind if you try "diagramming" the flow of your freely associated ideas, as illustrated in the two samples.

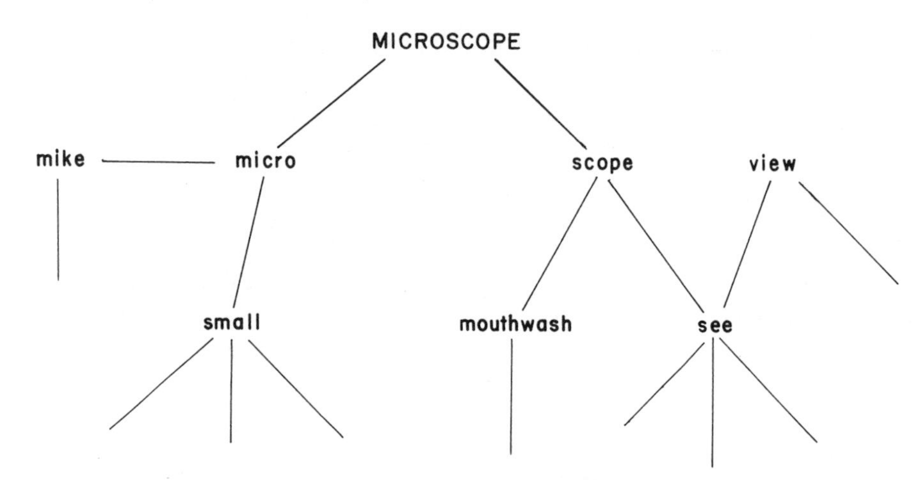

Fig. 7-3. Try your own hand and mind at extending free-association concepts from the single word, *microscope*.

Dosage 13: Concept association

This time, do a free association of concepts. Take a phrase of words that expresses an idea and write out any thoughts that come to mind, whether as single words or as full phrases. Your mind might play with the idea expressed in your original phrase, or it might play with the sound and meaning of a word itself. After you have listed the associated concepts, try putting some of them together in a paragraph. Suggested phrase: "Do it."

An example of a paragraph developed from a free association of concepts on "the letter A" follows.

A at once is seen to be acceptable as a letter of the alphabet, a word, a shape-indicator (A-frame), a sound symbol, number one to base 25, a superlative prefix (A-OK), or a subdivision of an outline (I.A.). As is true for most things in life, the symbol *A* has many interpretations in and out of context, enabling one to find a richness and a creative dimension in its use. A-frame up is a carpenter's delight but a defendant's plight.

THE PENNED TOOL

We hear, we speak, we push and pull information, at the start mostly by ear. Soon the subtleties of modulation become, unbeknown to self, the tricks of language. The way a thing is said becomes important to the understanding of its meaning. "The buck stops here; the doe goes into the larder." The continuity of sounds forms a comprehensible "put" into the mind. Habits of hearing, speaking, and writing become established. Unfortunately, no image from the conceiver can be put directly into the mind of another person. One man's tree is another man's home.

And yet, communication of information is so essential to the pursuit of life, liberty, and happiness, that to bridge the internal and external realities, humans have for milennia devised and practiced a simple-compound-complex system: language. As thoughts and needs developed, language exploded, multiplying sounds and sense, which in turn, by cell division, mitosis, or mutation (if not by proliferation and affectation), expressed even more concepts. Finally, the system got so big and widespread that, to facilitate its comprehension and encourage its use, folks began to talk about language itself. They began to describe language according to the habits with which people used it. As a result, we have dictionaries and grammar, both of which are descriptive guidelines of how the majority of people who use a language actually use it. If the science writer wants his article to be read, then it must be written so that it is easily understood; good writing is the product of writing conventions that achieve this end without offending the reader.

Dictionaries

Dictionaries and grammar are tools for the person who has need of transmitting his thought images to the mind of another. Dictionaries and grammar are essentially *de*scriptive and should be used as *pre*-scriptive only insofar as they indicate the best ways, because they are the most common ways, of letting others understand your thoughts.

A dictionary is indispensable even to the most practiced and prolific writer. For the science writer, two dictionaries are needed—an English dictionary and a science or medical dictionary.

Please be aware that language lives and changes; therefore, it is imperative to have a *recent* dictionary, especially considering that lexicographers are conservative with their descriptions of language. You need not throw away that first dictionary you used as a child and are sentimentally attached to; but, do not use it—enshrine it.

Like the naked emperor's clothes, the obvious is not always so obvious, so a few comments on dictionary usage follow. Different dictionaries structure their entries differently. In listing several meanings for a word, one dictionary might give the most recent meaning first and the oldest last, whereas another might do just the opposite. If two spellings or pronunciations are given for a word, one is not necessarily preferred over the other. The Introduction in a dictionary describes the particular system used in that dictionary.

A perusal of dictionaries reveals that different dictionaries emphasize different aspects of words. One has more complete etymologies than another; one gives fuller definitions; one gives examples of word usage in phrases or sentences after the definition; another gives synonyms and antonyms and distinguishes their shades of meaning. Whatever your other preferences, we advise getting a dictionary of the *American* language because British and American usage and pronunciation differ somewhat. Among the best dictionaries of American English are *Webster's New World Dictionary, The American Heritage Dictionary*, Funk and Wagnall's *College Dictionary*, and the Random House *College Dictionary*.

For insight into the heated (believe it or not) dispute over what a dictionary should and should not do, see "But What's a Dictionary For?" (*The Atlantic Monthly*, May 1962), which is Bergen Evans's answer to Wilson Follett's proclamation in "Sabotage in Springfield"

(*The Atlantic Monthly*, 1961) that *Webster's Third New International Dictionary of the English Language* is a calamity of the decades.

To be assured that you understand your dictionary's system, take Dosage 14.

Dosage 14: Dictionary peculiarity

Answer in your journal the following questions about your dictionary.

(1) Are the definitions of a word listed from most recent to most remote, or vice versa?
(2) Is the first pronunciation the preferred?
(3) Is the first spelling the preferred?
(4) Is the British spelling given?
(5) Are abbreviations such as "obs.," "sl.," "r.," and "colloq." used? What do they mean? What other abbreviations are used?
(6) When was the dictionary first published? When was it last revised?
(7) Are foreign words included?
(8) If a foreign word has become accepted as an English word and no longer needs to be italicized in writing, how is it entered?
(9) What is the etymology of *incisive?*

Producing a medical or scientific dictionary is a tedious, expensive process. Each publisher involved in that process relies heavily on the previous edition of his dictionary in preparing a "new" edition. Fortunately for publishers and unfortunately for the users, no one medical dictionary seems to meet all the criteria one would expect from a scientific work. Some give phonetic pronunciations for each word, others do not. Tables, charts, illustrations, special appendixes, and special word lists are common variables among the dictionaries. Some contents—for example, tables on modern drugs and dosages—are better sought in texts on the specific subjects. Typography that is declared modern and readable by one publisher is disdained by others.

Many factors contribute to the success or acceptance of a medical dictionary, but few actions seem to have more influence on that success than a favorable review written by a notable person and published in a prominent place. Notoriety of the editorial board also helps lay the foundation for success, giving a dictionary the proper "authoritative" input, personality, and credibility. Diversity of specialization among the board members is important, and diversity of board members' affiliations among commercial, educational, and governmental institutions or organizations is also important. Your choice of the most reliable dictionary in your field or specialty should take these factors into account.

8. Diction: You Write With What Kinds of Words?

Difficulties in scientific communications arise from several sources: the media condition us in specific word and phrase usage; our peers echo twists of phrases or nonces that "catch the ear" but become less disparate in repetition; the educational process demands that you create, scrutinize, evaluate, judge, think—except that formula-following, stencilling, and patterning belie the objectives of the process. Professionalism in writing carries with it the obligation to care for the language and its uses. Subsumed in that obligation is the need to be aware of the proper use of words.

BLACK TIE AND SNEAKERS

Diction is typically defined as word "level." Rather than rank words by such an unspecified ethic, we prefer to think of diction as word appropriateness. The appropriate diction will depend on a combination of person speaking, person spoken to, subject matter, and circumstances. Compare, or rather contrast, for instance, the language of the Nixon tapes with that of any Nixon TV speech.

As an exercise in awareness or word appropriateness, enter Dosage 15 in your journal.

Dosage 15: Who is your audience?

Do either of the following activities:

(1) Consider yourself the private physician of the governor of your state. You have made a diagnosis of the disease afflicting the governor. The disease is not one he would want broadcast. Write the words you would use to explain the diagnosis to the following people:

(a) the governor;
(b) a consulting physician expert in the medical specialty, whom you are meeting for the first time;
(c) a second consulting expert, an associate physician with whom you were in medical school, residency, and partnership practice;
(d) the governor's spouse.

The language should be appropriate to the audience and to the physician speaking.

(2) You are a biochemist, anatomist, neurophysiologist, or some other researcher in the sciences. You have just proved your hypothesis through laboratory experimentation. Explain the essential element of that hypothesis to:

(a) your assistant lab technician;
(b) the head of your department;
(c) a colleague visiting from Switzerland;
(d) a class of freshman biology or medical students;
(e) an audience at a meeting of your specialty's society;
(f) your spouse.

Some dictionaries give cues to word appropriateness by flagging words as informal, colloquial, obsolete, or vulgar. Dictionaries vary in their use and meaning of labels. Dosage 14 should have made you knowledgeable about your dictionary's definition for these terms.

The basic categories of diction can be termed formal English, standard English, colloquial English, and slang. Formal English is known as high falutin' by most people in the United States. Because Americans tend to be casual in all life habits, they do not use formal English much. It has occasionally characterized the style of serious scholars, but their names tend not to flip off the tip of one's tongue. Our presidents and other politicians certainly do not use formal English—they use a peculiar mutation, what Pei calls "double speak."* (In olden times, this might have required a forked tongue; today it is merely required.) Contemporary novelists do not seem to need or use formal English to maintain their image as intellectuals; most formal words have more than four letters.

*Mario Pei, *Double-Speak in America.* New York: Hawthorn Books, Inc., 1973.

Fig. 8-1. Meally worm with forked tongue, at the ready.

Inappropriately, certain groups of writers, science writers for example, strive for formal English; feeling somewhat out of their element, they are overly eager to be proper. They come to the picnic in tuxedoes, a bit like the "nouveaux riches" Balzac might describe. Consequently, in an understandable attempt to sound intellectual (instead of just intelligent), they write "utilize" or "employ" instead of "use," "disorganized" instead of "unorganized," "disinterested" instead of "uninterested," "interface" instead of "interact," and "methodology" instead of "methods," without realizing that their chosen words malapropically do not mean what they would like them to mean. (Parenthetically we add that, of course, if the misuse continues to be propagated and the majority conforms, then those words will *change* meaning; however, at that stage of development they will be so common that writers will no longer feel intellectual in using them and will probably corrupt some other like-sounding word—they might even revert to the original word!)

An enjoyable way of learning the origin of a word is to study its etymology, which is its etiology and natural history. (Incidentally, a side effect of studying etymologies is the clarification of the meaning of words.) Dictionaries provide the etymology or root derivation of a word, usually in parentheses, either before or after its definition. Some dictionaries are more thorough than others about etymologies. In particular we enjoy using *Webster's New World Dictionary*, in which "Etymology has been made a strong feature . . . because it is believed that insights into the current usage of a word can be gained from a full knowledge of the word's history. Particular attention is paid to showing these relationships as fully as possible and to carrying the etymologies back where possible to the Indo-European base."*

Enjoyable to read is Funk's *Word Origins and Their Romantic Stories.*** You might re-view yourself as writer if you know that "author" originally meant "one who originates something, makes something grow," and you might work differently with your "text" if you see it as "something woven." (We spin a yarn and weave our words and sentences into a fabric.) Writing faults might be clearer if you understand "succinct" to come from the Latin "succingere" meaning to gird up or tuck up short, "trite" to mean shopworn from too much "rubbing," "subtle" to mean finespun ("sub"=beneath, "tela"=web), and "plagiarism" to derive from "kidnapping." A "rejected" manuscript is, of course, one "thrown back" at the author.

Dosage 16: Etymologically speaking . . .

Look up the following words in your dictionary or another source and note their origins in your journal.

academy	doctor	panacea
adipose	elixir	quinsy
boudoir	labor	pundit
dean	nausea	X-ray

---

*"Guide to the Use of the Dictionary," *Webster's New World Dictionary of the American Language*, 2nd College Edition. © 1978 by William Collins & World Publishing Co., Inc.
**Wilfred Funk, *Word Origins and Their Romantic Stories*. New York: Funk and Wagnalls, 1950; paperback edition, 1968.

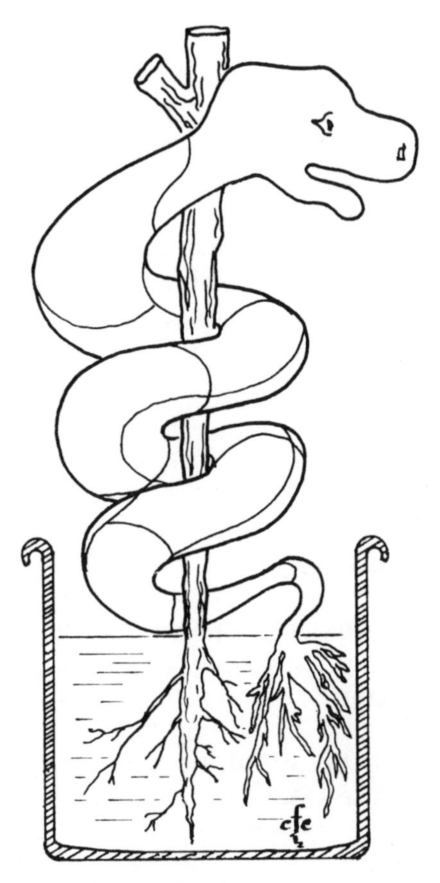

Fig. 8-2. Creative rootery.

STANDARD ENGLISH

The careful writer should know that readers are conditioned to certain expectations within formal articles, to other expectations within editorials and news items, and, of course, to directness in verbatim anecdotes. "Running amuck in the wards" may well be acceptable in reporting the patient's account of the scene, but elsewhere the editors would object to such subjective language describing the scene. The writer may be asked, graciously, to emend the description, whereby he resorts to an actual description of what the patients did; or, if a general statement suffices, he brings in "ward unrest," "patients' show of displeasure," or something similar.

What diction is appropriate for science writing? Standard English, the English most educated people speak, write, understand, and accept seems to be the most appropriate for science writing. Communicating well and quickly to a large audience shows intelligence and cleverness, and such communication is achieved through the sound symbols that are concise, precise, and familiar to your audience. Your message in science writing should be direct and in language that will endure. Use language that will be understood forty years from now; therefore, avoid slang, jargon, and transient expressions and write with an appreciation that, because of the ever-increasing multidisciplinary approach to science, those outside your discipline are likely to read your material.

Identifying the signs and symptoms of nonstandard English is one good diagnostic guideline for recognizing inappropriate word usage.

Anglo-Saxon or Latin?

English inherited, thanks to various conquerors and traders, words from several languages. The foundation language, Anglo-Saxon or Olde English, incorporated useful roots, words, and phrases, particularly from Roman and Christian Latin and Norman French. Indispensable as many of the words are, others are affectatious and entirely too long. Therefore, when you have the choice without changing meaning, use the Anglo-Saxon word instead of the Latinate word. Besides excising pretention, you will sever many syllables and streamline your message. Do you, for example, mean "death" when you say "mortality?"

For clues to identifying the origin of words by their prefixes, suffixes, and roots, consult an English handbook or study Latin, Greek, French, and German or Dutch. The following piles show a few options between words.

Latinate and Anglo-Saxon Piles

Latinate	Anglo-Saxon
acquire	get
alteration	change
approximately	about
avoirdupois	fat

Latinate	Anglo-Saxon
cessation, cease	stop
commence	start, begin
constitutes	is
corpulent	fat
demise	death
demonstrate	show
disclose	show
elevated	raised, high
hemorrhage (v.)	bleed
identical	same
inaugurate	start, begin
indicate	show
initiate	start, begin
inquire	ask
invariably	always
involuntarily contract	jerk
numerous	many
oral communication, parlance	speech
paramour	lover
performed	did, done
possess	have
present	show
prior to	before
provide	give
regarding	about
render	make
respire, inhale and exhale	breathe
reveal	show
similar (to)	alike (like)
sufficient	enough
termination	end

Slang (Today's Buzzwords)

Other types of words have entered the English language more recently than the Roman, Christian, and Norman Conquests. Clichés often began as slang words that originated as fresh twists of phrases, but became stale as they became popular. In the Charleston era, "cut a rug" meant dancing, presumably to give the image of frenetic movement tearing the rug to shreds. Slang expressions are usually first used by an "in-group" and then by a larger population. Slang is colorful, but often ambiguous—even to the in-group. In a freshman English class at a university, a student's opinion of a short story was that is was "freaky." Half the class understood "freaky" to mean "great," half to mean "terrible"; the instructor understood nothing by the term since the story was not gothic. Slang is a conversational and informal level of diction and has no place in standard science writing (though it might enliven a class lecture).

Neologisms (Tomorrow's Buzzwords)

Having gone through all the humor and lamentations about Pentagonese, Newspeak, and their products (counterproductiveness, benign neglect), we enter a new time for new words and new needs to attend to what goes on in the organicity of language. Undoubtedly the greatest influence on that organicity is advertising through the media, whereby great strides are taken to improve knowledge, as witness "quench-thirster" and the awesomeness of "static cling." Sports, too, contribute wondrously: "A-way to go!" "Good eye!" "Way to cut!" "Kill!" "Slaughter!" "Trample!" "Wipe out!" and the like.

If the time is right and presents an "opportunity," why not refer to it as an "opportunitime?" If the patient is more than ebullient, showing a character more like "bubbles sailing" than like one "boiling up," perhaps he (more likely she) is "ebbublient." One author who sailed his specimens from the operating room to the pathology lab via a guy wire called his technique the "Swish system." A few years ago, "prune-belly" became noted as a symptom, best described as such, obviously neologistically.

New processes demand new labels. Xerox, despite its campaign against such usage, will hear its process called "xeroxing," whether done on its own machines or on others. What else can we call a zipper, once a bona fide trade name? Will we be "fortraning?"

Authority gives one the privilege to coin new words. One may, in fact, wonder whether authority derives from such coinage. We are all aware that the water closet was invented by John Crapper. An assured road to fame or infamy or, indeed, chastisement is to put one's name to a process or product: Rex's hop-to test, August's time-boil analysis, Fernal's electrophoresis simplification. If fame is the directive force in your writing, look elsewhere than the bones, muscles, or other body parts. Your success will be more assured were you to correlate two or more previously uncorrelated symptoms/signs, whence the "syndrome" would become yours and your office staff would be kept busy with responding to requests for articles and speech-making.

Can you use "petri-dish," "input," "computer," and "digital" in a way that will make them and you famous?

Jargon (Professional Buzzwords)

One becomes steeped in the vernacular of one's discipline in the elbow-rub among peers, the wag of tongue at lunch, and the quick-read of articles in special journals. You, the scientist, learn the words and phrases that are idiosyncratic to your particular interests, those words and phrases serving as common code, identification, and communications milieu for the peer group.

A jargon word is an expression or a technical term used by a particular group or profession. Jargon is exclusive, often barbaric and imprecise. All professions have their jargons. Some jargon words are justified, especially in the sciences, where new advances call for new words. But most jargon terms are either clichés or barbarisms, that is, manglings of English due to ignorance of how language is commonly used. When a clinician asks, "How did the patient present?," the logic of grammar wants to know: "present what?" "She presented with headache and nausea." The verb "to present" standardly takes an object, not a preposition: She presented what or whom with a headache and nausea? What is meant, of course, is that the patient *had* certain symptoms or even (to stretch too far but not to mangle) that she presented certain symptoms to the physician. Probably the simplest way to report the case would be to say: "The patient's symptoms were a headache and nausea." Likewise, when a surgeon "biopsies a patient," he is using jargon. The *patient* does not undergo a biopsy—some part of his anatomy does.

Most jargon can and should be omitted from science writing not only because it is exclusive and often erroneous grammatically, but also because it is imprecise, and one sloppy thought can be the beginning of a pattern of sloppy thinking.

A list of common jargon terms in science and medicine is given below. Others are explained in the Glossary.

Jargon Pile

Jargon	Standard
diurese (a patient)	treat with diuresis
parameter	measurement, value, variable
patient presented with	patient had/showed; symptoms were
infant death	death of an infant
surgical intervention	operation
female, females	woman, women; girl, girls
male, males	man, men; boy, boys
differential factor	difference
neurological findings	signs, symptoms
operated a patient	operated on a patient
radical surgical attack	radical surgery
patients were followed	patients were observed
the art and science of medicine	medicine
patient revealed	patient showed
toxic patient	toxic reaction
surgeries	operations
downhill course	condition worsened
acute abdomen	acute condition of the abdomen
blood sugar	blood glucose
cardiac diet	diet for patient with cardiac disease (or specific kind of diet, e.g., low-fat diet)
clinical material	patients in the study
clinical situation	patient's condition
congenital heart	congenital heart defect
normal cytology	cells were normal
leukocytosis of 15,000	leukocytosis (15,000/mm^3)
patient's pathology	patient's disease
prepped	prepared
serology was negative	serum was normal
skull series	X-ray of the skull
urinary infection	urinary tract infection

Jargon	Standard
upper respiratory infection	upper respiratory tract infection
jugular ligation	jugular vein ligation
obesity pill	diet pill
had a fatal outcome	died
had a temperature	had a fever
autopsy examination	autopsy
bacteria exhibited growth	bacteria grew
clinical picture	observations
operated side	side operated on
discharged on a drug	the patient was discharged and given a regimen
cases	patients (where such is meant)
patient possessed	patient had
condition created	condition led to, precipitated
presenting symptoms	patient's symptoms
patient exhibited	patient showed, had
an alcoholic case	a case of alcoholism
expired, succumbed	died
patient was evaluated	patient's condition was evaluated
the literature	other publications, the science literature
differential count	differential cell count
the hemoglobin	the hemoglobin level
prothrombin	prothrombin time
sed rate	sedimentation rate
total cholesterol	total serum cholesterol value
animals were sacrificed	animals were killed (or state specific method of killing animals, e.g., were anesthetized, decapitated)

Dosage 17: Jargon heap

Entitle a page in your journal "Jargon Heap" and list as many jargon terms, with their standard variations, as come to mind that are not given in the jargon pile here. Then from the next five science articles you read, add all the jargon words you come upon.

9. Style: You Prefer Which Words?

The wonderful change in attitudes, evident in the past decade, is shown in the use of language. Today, anything goes. In fact, everything goes, including lucidity, comprehension, and downright meeting of communication needs—to the point where much is written but little is given. Erstwhile exiles—ambiguity, incongruity, pap, and circumlocution—easily and commonly reenter into language use.

Consider the wisdom and implications of Watson's comments in *The Double Helix* on the relationship between meaning and style and an author's personality.

> By the time I was back in Copenhagen, the journal containing Linus' article had arrived from the States. I quickly read it and immediately reread it. Most of the language was above me, and so I could only get a general impression of his argument. I had no way of judging whether it made sense. The only thing I was sure of was that it was written with style. A few days later the next issue of the journal arrived, this time containing seven more Pauling articles. Again the language was dazzling and full of rhetorical tricks. One article started with the phrase, "Collagen is a very interesting protein." It inspired me to compose opening lines of the paper I would write about DNA, if I solved its structure. A sentence like "Genes are interesting to geneticists" would distinguish my way of thought from Pauling's.*

VIGOROUS WORDS: VERB-SENSE MAKES SENSE

Some say that language is living; others show that it is dying. All presume that language began in the grunt-vocalizations of our simian ancestors. Great^n-grandfather Adamape looked about him and meta-

*James D. Watson, *The Double Helix.* New York: The American Library, Inc., 1969, p. 31.

grunted, "rok," "stik," "dog," "apple," and other things about objects that came into view, as is confirmed by the Bible without specification or refinement. Adamape, we are told, named everything, but it was Evape who consorted with the serpent to put action into the garden, whereafter eating, knowing, cursing, and being aware of nakedness gelled in their respective and collective minds and compelled them to be expelled from the passivity of Eden. (Rumor has it that the serpent still lives there, coiled on a modified Aesculapian Staff, repentant, contrite. We suspect otherwise.)

Had Adamape, like Evape, imitated his maker, we, their aftbears, would have inherited a pair of dominant genes for active communication in the manner of the maker who inspirited communications universally with such directives as "let be," "shall not eat," and, alas, "go forth!" Instead, we see that language began with a grunt unto noun-sense, proceeded toward a verb-sense, became entertainingly involved in the intellectual play of abstractions, peaked with the discovery of the Bic, and now, atavistically, is slipping again toward noun-sense and the ultimate grunt.

Nouns do predominate in our language today, and this is the pity because language is active, and activity pertains to the doing, moving, thinking, ebb-and-flow givens of communication. What can be said of a *thing*, when thing sets the limit of one's communication? The thing-defender may well retort, "Qualities and attributions . . . and such as that. *Red* thing, *big* thing, *protrusive* thing, *useful*, *necessary*, *natural*, *phenomenological* thing!" Does that convey the full awareness in representation, for example, of W. C. Field's nose, not just a nose, any nose, mind, but *his* nose, that brag-emblem, swash-bumbling, waving, twisting, almost-evanescing humor object?

Most science writing today is object/concept oriented: subjects, tests, methods, results, conclusions, references, hypotheses, theories, and . . . rejections. Even when what the writer reports centers on process and activity, reading and habit direct him to that noun-sense. Consequently, noun-use increasingly supplants verb-use, leaving few verbs to give spirit to or to convey the message.

Consider these examples:

The function for examination here is . . .
Freud made a significant contribution . . .

An additional consideration is Freud's conceptualization of . . .
Although the idea of death has metaphysical implications . . .
An attempt was made to develop a definition and description of . . .

Such structures fill the reader's mind-lobes with thing-words, bulges of collective masses, and bulks of burden. Action would invigorate the statements, help form the thought better, and thump the whole of it with spirit, movement, and process.

Is noun-sense, then, to be abhorred? No, nouns, obviously are necessary, but they should not substitute for verbs. The words listed here exemplify common nouns, rife in science publications, that apparently pop on paper as the writer's first-thought, really unthought, expression:

utilization
manifestation
development
identification
demonstration
consideration
illustration
separation

Note that these words are Latinate, longer than a verb replacement would be, and require at least one other word ("of") to make them useful.

Dosage 18: Verb-sense

Convert the windy object/concepts listed above to show action. Try to use as short a verb as possible, if possible an Anglo-Saxon form instead of a Latin form. For example, do not replace "demonstration" with "demonstrate," but with "show." If in doubt, consult your dictionary. Find ten more noun-sense words or phrases in an article you are reading or writing and convert them to vigorous verbs. You may have to rewrite the sentences.

This is not to say that all verbs invigorate language. The hack marks on the cave walls at the source of d'Nile have been interpreted: HIBI HUBI DUBIUS. Authorities attribute the ubiquity of the verb *to be* to that source–thence is borne our burden to be.

There *is* a tavern in the town.
The town *has* a tavern.
The town *boasts* a tavern.
Town and tavern *stand* atwain.

Where would science writing be today without the following spiritless sanked-'ems?

effect	do
present	take
undergo	have
serve	occur
perform	make

Watch for such words; often they merely clutter sentences.

Cluttered: Dr. Warren performed the operation.
Direct: Dr. Warren operated.
Cluttered: The experiment served to settle the question.
Direct: The experiment settled the question.

Here, a brief commentary on the word *virus* illustrates the advantage of verbs more vigorous than those ordinarily seen in science writing. Read Paragraph A first, then Paragraph B (an example of the usual fare). Then reread Paragraph A and do Dosage 19.

Paragraph A

The ancient Romans *originated* the word "virus," which for them *meant* a noxious agent or poison. Later the word *connoted* any infectious microbe, whatever its characteristics. Its meaning *was refined* only after successful research on treatment of diseases whose agents *could not be seen.* Jenner, in 1798, first *practiced* vaccination by using cowpox-lesion exudate against smallpox, and Pasteur, in 1884, *developed* a rabies vaccine; but, neither man *recognized* the specific cause of the disease he *was treating*. Investigators seeking that cause *knew* it *could result* from several agents such as poisons, toxins, and pathogenic bacteria. Meanwhile, Koch, Pasteur, and Ehrlich *isolated*, *described*, and *cultured* bacteria and *proved* specific ones pathogenic. They *showed* that some material—other than bacteria, toxins, or poisons—*remained* infectious after *passing* through bacteriological filters and *caused* many diseases. Eventually, such agents *were named* "filterable viruses" and later, simply "viruses."

Paragraph B

The word "virus" was originated by the ancient Romans and was considered a noxious agent or poison. Later the word was used to indicate an infectious microbe, whatever characteristics it had. Its meaning was more clearly defined only after research on diseases whose agents could not be seen developed successful treatment. Vaccination was first used by Jenner in 1798 when he employed cowpox-lesion exudate against smallpox. In 1884, a rabies vaccine was developed by Pasteur. Neither man was aware of the specific causes of the disease he was investigating. Researchers studying the cause believed it could be caused by several agents, such as poisons, toxins, and pathogenic bacteria. Meanwhile, bacteria were isolated, described, and grown in culture by Koch, Pasteur, and Ehrlich, and specific ones were demonstrated to be pathogenic. They also demonstrated that the etiological factors of many diseases were not bacteria, toxins, or poisons, but material that was infectious after being passed through bacteriological filters. Eventually, such agents came to be known as "filterable viruses" and later, simply "viruses."

Dosage 19: Vigorous verbs

(1) Reduce these cluttered sentences to direct statements:

When you do your exercises, take a firm grip on the lever that serves to activate the winch.

New liver cells are formed to take the place of those that are lost.

In the event that you cannot make enough observations of the timed trials, you should effect another selection of random subjects.

How much time does it take to make the animal respond to the stimulus?

About 16 million Americans are estimated to have gallstones.

She performed the total left hip replacement using the new prosthetic device.

The diagnosis of chronic gout was made.

The overall goal of the study was the development of a method of modification of human malignant cells.

(2) Introduce more vigorous verbs into this paragraph.

The data on complications of endoscopy *are included* in Table 1. Different complications occurred with different techniques. There *are* three methods now *employed* for treatment involving esophageal dilation. These *are* mercury-filled rubber bougies, metal olives passed over a guide-wire or string, and pneumatic balloon dilation. There *were* no statistically significant differences in complication rate between the instruments *produced by* different manufacturers when the instrument *was designed* to *perform* the same procedure.

CONCISE, ECONOMIC WORDS

Show me a writer and I will show you one who likes to prattle. Pretend you overheard that at a convention, because 'tis true, the fact of overhearing aside. Writers do enjoy giving importance to inanities and trivia, and often only the "self" or "style" claimed to reflect the writer in his work are what remain in the final published form, and pity the poor reader in search of substance.

Standard English is concise and economic. Be specific. Be brief. For other forms of writing, proliferation of words may be a virtue; however, for to-the-point science articles, the idea should be thoroughly expressed by brief, concise, and precise words. If you know your words and their exact meanings, you need only one word for one thought unit.

In a long letter to a friend, Ben Franklin wrote, "Had I had more time, I could have written it shorter." Indeed, in our haste to spill our currency on the pages, we cause readers to bear more than is necessary. It is more economical for the writer to spend more thought and time on the message to write it shorter than for thousands of readers to spend time deciphering it.

The Debris Pile below lists words and phrases that can be reduced to a simpler form or a single word because half the words are unnecessary to the concept. In some of the expressions, the words are redundant. Study the examples listed, then take Dosage 20.

Debris Pile

Wordy (Redundant) Expression	Concise Expression
quite full	full
past history	history
mortality rate	mortality
morbidity rate	morbidity
general public	public
joined together	joined
past record	record
past experience	experience
experienced a weight loss	lost weight
extremely likely	likely
one-half	half
grainy in character	grainy
oval in shape	oval
produced an inhibitory effect	inhibited
an order of magnitude greater than	greater than
in my opinion, I think	I think
sigmoid colon	sigmoid
two-week period of time	two weeks
interim of time	interim
rarely ever	rarely
allergic sensitivity	allergy
tied ligatures	ligated
surgical procedure	surgery
duration of time	duration
consensus of opinion	consensus
fuse together	fuse
pink in color	pink

Wordy (Redundant) Expression	Concise Expression
in order to, for the purpose of	to
hard in consistency	hard
rough in texture	rough
sour tasting	sour
the author(s)	I(we)
course of recovery	recovery
(sickle cell anemia) is a familial disease	is hereditary
subsequent to	after
it is important to note that	note that (or delete)
present time	present
this point in time	now
circular	round
communication	note or message
utilize	use
endeavor, attempt	try
etiology, etiological factor	cause
evident	clear
extremity	limb
location	site
it has been shown that	Jones and Eppilito showed that
in a very real sense	(delete)
in all probability, likelihood	probably
in case	if
in excess of	over, more than
in number, in length, in size, in volume, in area, etc.	(delete)
in spite of the fact that	although
it is well known that	(delete)

Dosage 20: Pruning debris

Reduce each of these terms to just a word or two:

it is obvious that
in order to continue
in the middle of the afternoon
in the middle of the day
serves as an incubator

Reduce each of these sentences:

It is of interest to note that in these studies, the effect upon glucose was not mimicked by the exogenous administration of the drug.

There is reason to believe that the test apparatus may possibly have been defective.

We are of the opinion that the roughness of the texture and the scratchiness of the sound when the material was scraped with the thumbnail led to his distress.

By far the vast majority of physicians believes that the treatment is efficacious.

VAGUE WORDS (MEANINGLESSNESSENSE)

The word "vague" developed from the Latin word "vagus," meaning "wandering," and wander the reader's mind does, confusedly, when he reads unanchored filaments of non-ideas. And, thanks to advertising and politics, non-ideas are ubiquitous. If you read that a large number of congressmen favor appropriating money as temporary relief for families in economic plight, you do not know *how many* (a majority?) of *which* congressmen (senators or representatives, Democrats or Republicans) or who (names, please) "favor" (a good idea but I will not vote for it?) *how much* (millions? billions?) by what *means* (loans? grants? gifts?) to *how many* (two hundred thousand? a million?) families (that leaves widows, other singles, and old folks out but includes the mafia?) for what *end* (a second car? food? shelter? education? travel allowance for escape to Mexico?) in what kind of plight (the plight of the consumer or big business families?).

Likewise, when the physician reports that the patient "is doing as well as can be expected," does that mean he is firming up as the mortician embalms him? And what about the "high" cholesterol values, not to mention the "significant" ones. We are being picky? That is right; would you want Dr. DeBakey to instruct you to insert a valve in the "vicinity" of the heart? The reader needs specifics. Science writing should report scientific data or knowledge. Defined in its broadest sense, science is "a possession of knowledge as distinguished from ignorance or misunderstanding." Vague and ambiguous words are imprecise and by definition have no home in science; they are justly doomed to wander about empty halls and wastebaskets, banned from laboratories and manuscripts. A reader cannot help but suspect that the scientist who reports "large increases" in triglycerides is measuring "large amounts, more or less" of serum in his medium-sized test tubes. And be careful about giving your patient a "recommended" small dosage of vasopressor. Many vague words try to describe size or magnitude of relationship to a standard; the actual values should be given.

Must every clinical or experimental situation be reported in the same way, with the same descriptors and phrases? How lost are the meanings of some words now because of their overuse? Where is the specificity in "demonstrate," "utilize," or even "discussions," words that are building elements of many scientific communications? If you do demonstrate, then, by the coiled Lapsus, state it thus; otherwise, consider that you only show, note, see, report, or record.

Review Chapters 5 and 6 and Dosages 5 through 9 in your journal to recall the reasons words can miscommunicate information, then take a large amount of Dosages 21 and 22.

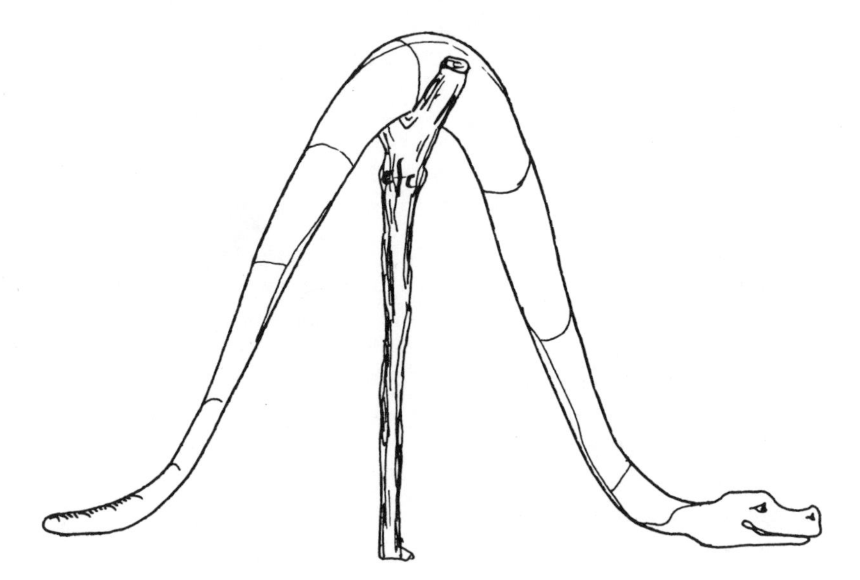

Fig. 9-1. Lapsus demonstrates normal distribution curve with mean (crotchety-cratchety) stick.

Dosage 21: With precision, please . . .

(1) Refer to your dictionary and write in your journal the specific differences among the following words: determine, analyze, evaluate, count, enumerate, discriminate, identify.
(2) Underline at least six vague concepts in the following sentence:

> As part of an ongoing interdisciplinary study we set out to measure the effects of exercise on obese persons attempting to reduce by dieting.

To become more familiar with the fine distinctions between words, read through the Glossary of this book (Chapter 21) and those in the recommended reference books (Chapter 20).

Dosage 22: The vague nerve of it . . .

Critically review several pages of one of your articles and of someone else's and list the vague terms. The "Introduction" is usually a Pandora's box, so start there. Add your words to the list below.

Vague Word Pile

he was real good
it was great
a vast difference
of grave concern
a sizable increase
doing as well as can be expected
of no avail
few and far between
great amount
significant periods of time,
  for a long time
very large
extremely old
little opportunity/great
  opportunity
a somewhat high complication rate
heavily contaminated
several
frequently
thoroughly
declined
closely
severe
underlying
most

very high
very rare
vast majority
valid reasons
grand and lofty visions
tight budget
large classes
limited time
quite well established
imaginative design
a creative experience
meaningful and worthwhile
  learning experience
well illustrated
several years ago
the past decades
recent decades
prodigious efforts
key areas
lump sums
changed markedly
maximize one's options
more infrequently
allow nature to take its course
needless to say
a lot more
reliable sources
first-rate
top form
in his prime
means to an end
precious little
considerable increase
considerable difference
significant change
significantly decreased

suggested
seldom
most probable
overall decline
clearly reasonable
containing large numbers
practically ceased
rigorous
significant
extremely
increased risk
almost all
clearly defined
for a long time
considerable complications
uncommon
progressive increase
majority
much lesser extent
roughly parallel
disproportionate number
greater
had recently undergone
valuable insights
extensive literature
the literature
numerous studies
appears that
small discontinuities
frequently observed
at times
potentially lethal thrombus
threaten migration
several months history
sublethal injury
massive or submassive
  pulmonary emboli

CLICHÉS

Certain writing faults will annoy the reader because they keep him from getting the point quickly and clearly; they waste his valuable time. Among these faults are imprecise words (either vague or ambiguous), redundancies, jargon, and clichés.

Standard English avoids clichés. Words that repeatedly cluster together in speech and in writing are called a cliché. Given the first word or two, one can predict the rest of the phrase or expression. Fill in the blanks to see for yourself:

right as ________
strong as an ________
jumped for ________
abreast ____ ____ ____
conspicuous by ____ ____
last but ____ ____

Or you may prefer the more difficult form of the question-and-answer quiz:

Where does one *hit the nail?* ___ ___ ___
What kind of *example?* ___ _____
How *sure* are you? ________
What can't you pull *over his eyes?* ____ ____
You are *as free as* which creature? ___ _____
How *sly?* ___ ___ ____
How *quick?* ___ ______
What about him do you *hate?* ____ ____

Because clichés are common, they should result in easy understanding, one might argue; therefore, why not encourage their usage? Occasionally, the cliché is the best form of expression, but such occasions are rare and usually involve special objectives, for example, humor or emphasis of the familiar. Art Buchwald has a lot to say about clichés.

The cliché comes easily to mind, particularly in the course of conversation with a friend. The easy-come is the first objection to the use of clichés; the writer may be comfortable using them and find his writing flow eased onto the paper, but the reader is likely to find clichés imprecise and boring. If you're quite sure you can't pull the wool over Bert's eyes because he's the kind of guy who'll hit the nail on the head every time, what do you mean exactly? Reliable Bert can't be fooled.

Another objection to the use of clichés is their commonality and lack of richness; the reader need not be entertained or find poetry in every line of writing, but the property of "you-ness" or originality in the writing should be evident and interesting to him. If Gert, who is strong as an ox, free as a bird, and sly as a fox, is conspicuous by her absence, would you jump for joy? She is, after all, the perfect example of someone who does not keep abreast of the times, proving ignorance is bliss, and you hate her guts.

Quick as a rabbit, the reader knows what you are trying to communicate, right? Maybe, only maybe. Imagine the befuddlement of a youngster growing up in Texas who hears poetic references to the strength of the mighty oak—when all he sees are prairie scrub oaks. What do all the clichés on strength, freedom, slyness, and ignorance convey to the reader? Primarily that the writer did not know how else to express the thought and, hence, is not much of a writer.

Treat yourself to a thirty-minute television program one evening, and along with it take Dosage 23.

Dosage 23: TV cliché

Watch a thirty-minute television program with your journal and pen in hand. Record in your journal each cliché that you notice in the dialogue or in the action. (Better have a fast-rolling ballpoint.)

A cliché was at one time an original, probably perky, way of describing an object, action, or quality. However, use and overuse of the words have rendered the fresh image stale and worn out; the image is dead. The cliché comes easily to the lazy mind; one need not think—neither writer nor reader—and therefore no one shares a new idea. Besides, clichés require too many words for too little thought; so, who needs it, who reads it?

As an exercise in spotting clichés, read the examples given below; then take Dosage 24.

The Petty Etiquette Survey

Imagine finding yourself in the following situation. Select the comments you would like to make; then select the answer that is a cliché.

(1) While fishing, you meet the ghost of Isaac Walton, author of *The Compleat Angler*, wearing panache and codpiece.

(a) I don't know anything at all about angles.
(b) That's some fly.
(c) As a baiter, you must be a master.
(d) I've always wanted to have the opportunity of meeting you.

(2) On encountering a local politician in your psychiatrist's office:

(a) So this is where you learn to couch your terms.
(b) Obviously you are a victim of the couch paradox, which enables one to tell the truth whilst lying.
(c) Your mind to shrink, your head to cloud, your hand steady on the till.
(d) Excuse me.

(3) Finding a gold coin on the sidewalk outside your bank and grabbing it simultaneously with an attractive member of the opposite sex:

(a) Heads, I win.
(b) Tails, I win.
(c) Aha, you found my guilt-edged King Kluup Chronic Wonder Wander Kroner . . . and I thank you.
(d) Let me help you with that.

(4) At the airport, you lose your luggage only to find it being rifled by a famous movie star:

(a) Mother always said I would one day lose my grip.
(b) There must be some mistake.
(c) Lights . . . Camera . . .
(d) You won't find it in there, Inspector, for Bogart smoked it in the last reel.

Dosage 24: The petty etiquette survey

In your journal, record at least four "appropriate responses" for each of the following situations:

(1) An excited physicist showing you his tracings as proof of discovery of a new particle.
(2) In "closing," you have sewn your glove to the patient and then turn to the chief of the surgery department, saying . . .
(3) Appearing before the Senate Subcommittee on Investigation of Federal Gum (or Gun) Control, you begin . . .
(4) A colleague notices the fuchsin stain on your lab coat, just below the left elbow.

Now, invent several situations yourself.

Can you bear a little self-criticism? Take Dosage 25.

Dosage 25: Cliché!

Take an already published article of your own *or* any other piece you have written, *and* another article in a specialty journal. Go through both and mark the clichés that are characteristic of science writing. Then rewrite them, de-clichéing them. After you have done this, you may glance at the list of clichés common in science writing provided below. Add your own discoveries to the list.

Cliché Pile

at the height of acute renal failure
corrected only partially
prominent among these conditions
which may account for
was no longer present
at the present time

Cliché Pile (Continued)

it is of interest
could be explained by
do not allow clear distinction
you can believe it
highly unresponsible action
others have already said
a man who has everything
no guide to the future
lose his shirt
smacks more of astrology
short end of the stick
put this thing to a stop
make do or do without
it's more than just an inconvenience
leading cause of death
heart disease is the number one killer
therapeutic advances
drug of choice
purpose of the study
therapeutic modality (!)
open and shut case
give it my best shot
in the present study we investigated the effect of
it has been shown that
accelerated coronary heart disease
there is evidence to suggest
objective measurement
is receiving increased attention
rule out
increasing evidence
preliminary observations
under the conditions of your experiment
seems to indicate the possible presence
neither fish nor fowl

SLANTED WORDS

The purpose of communication, especially in the sciences, can be further distorted by "charged" or slanted writing. Compare the following two lists of definitions.

List 1–from "Definitions not found in Dorland's Medical Dictionary," by John H. Dromey*

Ascorbic (Vitamin C): "Faddy" acid
Bedsores: Berth marks
Biopsy: A slice of life
Chromosomal change resulting from drug use: High-gene
Contraceptive: Labor-saving device
Diagnostician: Wizard of "ah's"
Discomforts of pregnancy: Heir sickness
General anesthesia: Technical knockout
Hyperactive child study: Speediatrics
Missing sponge: Gauze for alarm
Muscle: Cramp site
Nurse shortage: R. N. deficiency
Open-and-shut case: Exploratory surgery
Plastic surgery: Arts and grafts
Scheduled Caesarean sections: No uncertain terms
Tapeworm: Nibbling rivalry
Varicose vein: Lame duct

List 2–from *The Enlarged Devil's Dictionary* by Ambrose Bierce**

ABDOMEN, A shrine enclosing the object of man's sincerest devotion.
ADAM'S APPLE, A protuberance in the throat of a man, thoughtfully provided by Nature to keep the rope in place.
BIRTH, The first and direst of all disasters.
BRAIN, An apparatus with which we think that we think.
CORPSE, A person who manifests the highest possible degree of indifference that is consistent with a civil regard for the solicitude of others.
DISEASE, Nature's endowment of medical schools.
DOCTOR, A gentleman who thrives upon disease and dies of health.
EPIDEMIC, A disease having a sociable turn and few problems.
ESOPHAGUS, That portion of the alimentary canal that lies between pleasure and business.

*John H. Dromey, *Medical Opinion*, August 1973, p. 61.
**Ambrose Bierce, *The Enlarged Devil's Dictionary*, edited by Ernest J. Hopkins. New York: Doubleday, 1967.

HELL, The residence of the late Dr. Noah Webster, dictionary-maker.
HOSPITAL, A place where the sick generally obtain two kinds of treatment—medical by the doctor and inhuman by the superintendent.
PRESCRIPTION, A physician's guess at what will best prolong the situation with least harm to the patient.
QUACK, A murderer without a license.

How do Dromey's purpose and style differ from Bierce's? Does a specific tone dominate each list? Is the author humorous, satirical, or cynical? Dromey's definitions are humorous and, for the most part, they are puns. His images or metaphors (implied comparisons) are fresh—though if used repeatedly they would become clichés. Bierce's definitions are not puns, but are interpretations expressing a cynical point of view. As a nineteenth century American satirist, Bierce may have been trying to improve the quality of institutions and of man's ethics by using a twist of humor to point out the evils. At any rate, his words are emotionally slanted, and obviously so.

But the emperor's clothes are not always so obvious. A politician's speech or a news story may likewise be slanted; the full message comes from "between the lines," and the public's patterns of behavior may be shaped thereby. Today we have the "new journalism," although its newness has faded and it is truly "knew journalism." Stories written in that style are evinced from the mind and heart of the writer who, after all, is the final authority on what he is about to report. He writes what he *knows;* the trouble is, that is not the "nose for news" one hears of among the pundits. Objectivity is still an important part of getting information transmitted to the public. The "new journalist" may say he "knew" the truth of a story, but what he reports may be what he "sensed" was the truth of it.

The scientist may be reporting "new science" in his article but, excited as he may be about his discovery and as desirous of convincing others he is right, he should always maintain objectivity in his writing and avoid emotionally slanted language. If written clearly and to the point, his article will present the data convincingly enough.

The word "objectivity" is often used to justify a habit of science writers that new writers are repeatedly questioning—the omission of

Royko

Magnificent Mick, Super Stones Do Really Big Show in Chicago

By MIKE ROYKO

CHICAGO — I've listened to some pretty good music over the years.

When I was young, I spent time near New Orleans and heard the old originals play jazz. Later there was Louie Armstrong, Count Basie, Duke Ellington, the Modern Jazz Quartet, George Shearing. And Reiner and Solti conducting the Chicago Symphony Orchestra in Beethoven, Mozart and others.

At the time I thought I was hearing the best. Now I know I was wrong.

Possibly the greatest musical event of our time, or any other time, took place in Chicago only a week ago.

It was the performance by the Rolling Stones, a group from England that does music in a style adapted from the black-American blues.

It was my bad luck to have missed it, but there is no question in my mind as to the greatness of this group and of their Chicago concert. Not after reading a review by a rock critic that appeared in another Chicago paper. In case you missed the review and-or concert, here is what the critic said in his very first paragraph:

"How do you write with a blown mind? How do you collect yourself when your body and soul have been shattered? How do you tell someone about rock and roll? How do you tell someone about the Rolling Stones?"

I've read reviews of many fine musical performances, but this is the first time that a critic has pleaded physical and mental disability due to the magnificence of what he heard.

But fortunately this fellow recovered sufficiently to go on, blown mind and all, to tell us that...

"It's no use. Simply saying that the show the Stones put on in the Stadium Thursday night was electrifying, devastating and unbelievable doesn't do it."

That must be terribly frustrating, to hear something so good that merely saying it is "electrifying, devastating and unbelievable" doesn't get your message across. It's that kind of torment that causes extremely sensitive people to jump out of windows, or at least cut off an ear. But he tried to put it into words anyway.

MICK JAGGER
Gimmicks part of act.

"Picking out the finer points of the music that they made seems as irrelevant as timing the shivers of excitement they elicited. All seems to dim in their awesomeness."

I hope you appreciate what that seems to mean: The Stones are such an amazingly great musical group that even their music pales in the presence of their greatness. Nobody has ever been that awesome before. As good as Beethoven was, it was his music, not his bushy hair and eyebrows, that people dug. And nobody went into physical collapse over the handkerchief Louie Armstrong had to dab his lips, or even his lips themselves.

The critic then detailed some of the Stones' awesome qualities that drove him to the threshold of the emergency ward.

"The staging was superlative. The six-pointed star that opened up . . . was simply the best possible vehicle for Mick Jagger, the human hurricane. He used the star as six runways into the audience, not really for intimacy, because the Stones are never intimate, but to magnify his open-mouthed grimaces, taunts and provocations..."

"The gimmicks were there, too: the phallic balloon that appropriately climaxed "Star Star," the rope which Jagger swung Tarzan-like into the audience, the now-beloved belt-whipping of 'Midnight Rambler.'"

To think that a music-lover like me missed all that — an open-mouthed Mick Jagger charging into the audience to taunt and provoke, or swinging from a rope like Tarzan. A phallic balloon and the "now-beloved" belt-whipping. It would have been something to have told my grandchildren about, that's for sure.

As for the music itself, the critic said: "The numbers were a mix of old and new Stones classics and some oddities..."

That tells you how great they are. They not only have old classics, but they have "new" classics. I suppose that even now they are planning some "forthcoming" classics.

If you aren't yet convinced that this was the super-event, I have conclusive evidence.

In that same paper, only a few pages away, was another review. The paper's jazz critic wrote about Benny Goodman, 66, the semiretired jazz clarinetist, who came into town with a group of jazz all-stars.

"His tone remains pure . . . his intonation is always right on the button, and his inventions sound fresh to the ears. He remains the finest technician that jazz ever nurtured . .

Long runs are always logical and only occasionally surprising; the only "bluish" notes in his bag of tricks are still the minor third and seventh . . . Benny established and perfected a style years ago, and although other musicians have grown in many other harmonic and rhythmic directions since that time, Goodman has not. He doesn't really need to, I guess. His style is classic."

See what I mean? Not a word about Goodman being awesome, or devastating or unbelievable. Goodman didn't even shatter the critic's body.

That ought to tell you how much greater Jagger is, whether he is dangling from a rope, swatting somebody with a belt, or just taunting his audience with his girl-boy body, and mascara-smeared face.

And I'm sure he'll be just as great 30 years from now, when he has been around as long as Benny Goodman.

I can see it now — a 66-year-old Mick Jagger swinging into the audience, Tarzan-like, and the people excitedly shouting:

"Who's the crazy old lady?"

Figure 9-2. Reprinted with permission of Mike Royko and *The Chicago Daily News*.

the first person pronouns "I" and "we." An argument against the pronoun is constructed on the basis that science is a vast body of objective reality that cannot belong to individuals but can only be described by scientists; therefore, an author (the reasoning continues) should disappear from his article (even though there are diseases, syndromes, and bacteria named for his discoveries), unless, of course, the lines of the page take it upon themselves to refer to him as "the author." Bierce might typically define the avoidance of the first person pronoun as the fear of or refusal to take responsibility for having said what is said in the article. The argument for objectivity is self-contradictory, for what could be more objective than to clearly establish which bits of information in the article have been observed by others and which by the author, and which conclusions and hypotheses are those of the author?

For a summary-critique-example of all the smarting blisters of barbaric diction, consult Royko's article on an article.

SPECIAL WORDS

Idioms

One's ear is important to one's use of language. Sometimes word combinations belie logic; but, having learned a combination in a given context, the user knows that others familiar with the language will understand the meaning beyond the literal words. "Catching cold" is what everyone who has one does, yet logically no one really catches the cold, although few would state with assurance just what one does do.

The idiom in science writing is less frequent than in other types of writing, but the scrupulous writer should be aware of the form. Many idioms are formed with common verbs to which prepositions are added. Thus, one can:

go over (a test with a student/the boss's head)
have it out (with your spouse)
go on (TV, radio, the road)
turn over (your authority, a new leaf, in your grave)
set your heart on (a new piece of lab equipment)
keep at it (until the wee hours)
make believe (there are more than twenty-four hours in the day)
take on (new duties)

(note that she) takes after her mother (but that you do not)
turn on (your friend in anger, the radio, your then-to-be lover)

The caution about idioms is that the literal meaning may be taken when the idiomatic meaning is intended, especially, of course, among those for whom English is a second language.

Evolutionary Words (Neither Amoeba Nor Flagella)

Some words have entered the gray zone of change. Their acceptance hangs in the balance of editorial preference; but again, brevity, specificity, and clinical trends will push and pull them toward or away from frequency of use, for example, the use of "prepped" for prepared, and "X-ray" for both the roentgen ray and for the film made (the roentgenogram).

"Data," because it has a handy, though Latinate, form for the singular, "datum," has been considered to function as a plural noun and to require a plural verb form: the data are listed. Yet, most readers would place "data" in the same category of other collective nouns such as "group," "class," and "committee," whereby the need for a plural or singular verb form would depend on the context. "Data" is being used more and more in the same way as "information."

Often a noun has no equivalent in a verb form. Those whose activity is not limited by the formalism of proper use, joyfully, if indifferently, begin to use the noun in a "presumed" verb form. They may record that "the tissue was biopsied," not realizing that biopsy is supposed to be limited to use as a noun—at least in formal circles. Also, "they autopsied thirty-one patients" rings well in the ear of the researcher who did the retrospective study. Extending such terms as modifiers may qualify as abuse: "autopsied patients," "biopsied specimens."

What objection can there be to such use? Without the availability of verb forms, the writer is left with only a cumbersome structure on which to hang his thought:

The body underwent autopsy . . .
Specimens taken by biopsy . . .

Mention must be made of the word "control" as used in experiments. One controls the knob in a titration tube, the dial on a rheostat, the wheel of a car, and sometimes oneself. In experiments, however, the control group, oddly, is one that is not controlled;

rather, it is permitted to go about its business without interference from the circumstances of the experiment. Truly, it is the "uncontrolled" group in contrast to the experimental group, although, of course, both groups are part of the experiment. Confusing? Agreed! To replace the misused "control," "comparison group" should be introduced into trial use, and its legitimacy judged on its acceptance and appropriateness.

How does one treat such terms just entering the gray zone of change? In appreciating all of what has here been written, the writer should be aware that in using a singular with data, lab instead of laboratory, and autopsy as a verb, he may not be acknowledged as being *avant garde;* on the contrary, his less *au courant* colleagues might look on him as an unlearned simpleton, perhaps an unreading oaf. Nonetheless, formality continues to erode, and absolutes prescribed by editors will constantly be challenged by contemporary forces, forces greater than even the editors–money, materials, means, memberships, subscriptions–in the game of acceptability and legitimacy.

In the blacker zone of slang, vulgarities and obscenities had firmer fences built around their use in times gone by. Even the redoubtable *New England Journal of Medicine** recently published a comment on the question of the use of "fart" in preference to the longer, somewhat more cumbersome alternative of "flatus" (per anus having passed). Because many patients would better understand "fart" and because it is shorter, logic would lead one to suppose that it soon would become preferred. The vagaries of social pressure, the visceralness of the matter (!), and other natural positions on the subject must lead one to conclude that the four-letter-word (listed in *Webster's Eighth Collegiate*) must remain on bail pending further hearing of the evidence as to its innocence or guilt.

Chopped Words

Style continues to relax, and abbreviations now abound–to the point that their expanded terms are unfamiliar to readers, or at least not easily recalled.

*Robert J. L. Waugh, "Plain word for *passed flatus*," *New England J. Med.* (*Letter*) **296**: 178 (Jan. 20, 1977). Printed by permission.

Dosage 26: I used to know that.

Write out the full meaning of these abbreviations:

SGOT	DNA	pg	MHz
DMSO	MU		CFT
IUD	mRNA	PD	

Onomatopoeia (Simple Sounds)

Occasionally the writer faces the problem of having to convey to the reader an actual clinical sound. Lub-dub and its variants, saying "EEE" instead of "AHHHH," clucking, splatting, bow-wowing, and plop, plop, fizz, fizz–all relate to the use of words to convey the clinical sounds. Although such use occurs most often in fiction, poetry, and humor, the specific need for such use should oblige you to consider the effectiveness of an onomatopoeic word. What did the infant say, truly? The attentive writer will listen for special sounds that occur within his discipline: gurgles in laryngology, rumbles and spurtings in surgery, even the hum of the equipment and the vroom of the motorcycle of the free-wheeling researcher. Perhaps we have too much maligned the grunt.

Eponyms

Language meets its need, whatever the voice that may be raised to question its logic. As the Greats described conditions and structures and became identified with them, science accepted their labels: Parkinson's disease, Purkinje's cells, Von Linne's classification, Bergey's Manual of . . . . The popularity of such nominative activity led to some conditions of having regionally preferred surnames, other surnames having myriad conditions/signs/symptoms assigned to them, and taxonomists desperately seeking order out of chaos.

In review of the correctness of eponym use, the taxonomy and terminology hierarchies concluded that names should be retained in identifying diseases, syndromes, and signs as the most efficient and collective means of identification. Eponyms are now legitimate in specific uses, but they are not shown in the possessive form. One does not say "Washington's Street" and neither did Parkinson actually have the disease. Why show it as possessive? Ergo, Parkinson enters the scene, but without his apostrophe. In eponyms combining two or more names (for example, Guillain–Barré syndrome) the apostrophe was traditionally omitted anyway. Use of the possessive is even on the wane for such tenacious ones as Hodgkin('s) disease. An important reason for dropping the apostrophe in eponyms, however, is that it helps in spelling pertinent proper names correctly. Use of the possessive form of the eponym often causes surnames to be misspelled: Wilms tumor may well be found as Wilm's tumor. Other often misspelled names include Homans, Fuchs, Adams–Stokes, Calve–Perthes, Meigs, Mosse, Sturge, Thiele, and Danlos.

Dropping the possessive form and using a surname as an identifying label has the advantage of facilitating computer use. Early on, the need to put *Current Medical Information and Terminology** on a computer system necessitated either dropping apostrophes or overcoming confounding complications. Infinite wisdom aside, the confounding complication was not faced; the apostrophes were not included in the input and were thus excluded in the output.

*Burgess L. Gordon, (Editor), *Current Medical Information and Terminology*. Chicago: American Medical Association, 1971.

10. Word Functions: Piling Them Together (Or, What Is That Word *Doing* In Your Sentence?)

SIMPLE WORDS, SIMPLE COMBINATIONS

Were language, like gestures, limited to singular expressions such as shrugging the shoulders to show uncertainty, our tasks, our powers, and indeed our treaties would all be terse, ambiguous, and, by and large, meaningless. Apple. Eat-not. AHA. Go! Perhaps, after all, the Eden scene was an apotheosis of misunderstanding. Adamape might have turned to Evape to say, "I have a debble of a time understanding serpentalk. Waddidy say?"

Richness in language comes with variety in word usage, including word combinations that depend on position and context to set the meaning.

Simple Sense

To naively pursue a greater appreciation of use, relatedness, appropriateness, and meaning of words, you must know that the Latin for the four-letter-word "word" is *verbum*. Street-conversationalists and such four-letter-word notables as Noah Webster and Richard Nixon place those fame-words into contexts that connote action, represent objects, modify things and fellow personages, interject tone, and otherwise carry the rhythm, if not the substance, of language. In deference to those still distracted by such words, we opt here for a four-letter word that for most readers will rest on the sterile surface of indifference: i.e., suit.

A nice word is "suit." One may talk of his brown suit, zoot suit,

Sear's sucker suit, legal suit, suitcase, or (fancifully and for snoots) hotel suite. Try it on for size; enter here a word to describe suit in some way:

_______________ suit.

Thus have you specified the object meaning into something specific.

Now, to open the vista of word appreciation from noun-sense to verb-active verbatim, we must divest ourselves of the habit of considering *suit* merely as some *thing*. We may also find that something suits us, suits the occasion, or, in halcyon perspectives, suits one's fancy (suture-self, surgeon). With each different use, the word itself controls the meaning of other words.

Can every word be used in all forms? Fortunately for the masses, many words can; but, unfortunately, some words are forced into awkward situations. Consider these words:

test take today

Conceivably, you can test students, get test results, put them to the test, and fool around with testing, testers, retesting, detesting, pretesting, protesting, making the product testable, or even leaving your estate intestate.

Take "take":

What's the __________?	mis__________
__________ your time.	under__________
__________able	under__________er

Today forces the issue. Can you today something? Will you eat a today sandwich for lunch? Is the objective todayable? Is your best friend a todayer, a yesterdayer, or a tomorrower? One sees that today does not have the flexibility of the other words; it is limited today to its specific adverbial and noun meanings.

Pre- and Suf-fixing

Many words are composites in themselves and have origins in roots, and to those roots were attached beginnings and endings–prefixes and suffixes–to refine meaning.

Without even knowing whether a word derives from Anglo-Saxon, Greek, Latin, French, or American Indian you can take its basic form and maleate it, decorate it, and even give it a different architecture by adding accepted fragments.

Choose a word–pot, test, cupel–and we will proceed with first things first: prefixes.

Prefixes are first things, of a sort. They are usually short handles attached to the front of a word to twist, bend, enforce, extend, reverse, or otherwise direct its meaning. Hence, *pre-* means "before" or "in front of," showing that a prefix is something *fixed* before something else. Unfortunately, "fix" does not mean "after" or the whole question of additions before and after the word root could be resolved with the single form. To give symmetry to the concept, the language named the afterfixes "suffixes," probably because postfixes sounds too much akin to fencing, Washington journalism, or mail delays.

Common prefixes are *re-* (again, anew), *pre-* (already center-staged for you), and *com-/con-* (next to, with).

Try these in combination with your word choice:

pre__________
re__________
com__________

The organic quality of language is such that you may well find in your writing a need to form a new word to correctly and succinctly represent an observation. Facility with the interplay of these forms enables one to be more specific about meaning and message. Students become pretested, tested, and posttested. The instruments are the pretest, test, and posttest. Each is recalled by the students as a pretest, test, and posttest experience.

In using "pot" or "cupel" you may hesitate to form "repot," "precupel," or "postpot." Although the tools of language are there for all to use, they should not be ab-used. "Repot" has the danger of resembling despot, depot, impot(ent).

Consider the word "Pregame" used in a title in a medical article–"Approaches to Pregame Counselling." Many readers responded to that title with surprise in that they believed they would learn about a new condition, *pregame* (presumably pronounced preg-amy), when, on reading the article, they found that the comments were about athletics and how to advise participants *pre*-game. Ergo, care should be used in fixing new words. Why not say "Approaches to Counselling Before Games"?

In combining with some words, prefixes may visually convert the word's sound or meaning. "Preexist," when first seen, may cause the reader to "see" preex-ist rather than pre-exist. Conditioned by such words as preempt, reinforce, cooperate, and weest, and with the current trend toward abandoning hyphens, readers are now accepting such forms as reestablish, reevoke, and reinvolve. However, if combining a prefix to a word without a hyphen should confuse not only the sound but also the meaning, as in recollect (meaning collect again, not remember), the hyphen is warranted.

The prefix *bi-* is a double problem. How often does a biweekly journal get published? We know what a bicycle is, what biceps are, and who among our friends are bilingual. But what about those temporal confounds, bimonthly and biyearly? *Bi-* can mean "issued every two" or "issued twice during a." In forming words, consider what confusions may result, the delights of discovery in the new form notwithstanding—for example, your fellow worker who is consistently biassing his results, if not himself.

Policies among science journals differ for combinations with some prefixes—for example, *post.* Where one journal prefers post-traumatic, others opt for posttraumatic. Hence, the need for a good, solid noun caused one particular characteristic of the abnormal heart to be written arrhythmia, engendering a controversy about whether the correct form should be a-rhythmia, adrhythmia, dysrhythmia, or malrhythmia. We recommend that you rely on the dictionary and let the journals make their personalizations subsequently.

Dosage 27: Prefixing . . .

Choose four words: a noun, verb, adverb, and adjective. Then answer the following questions. (1) What does each of the following prefixes mean? (Your dictionary or your knowledge of Greek and Latin will help.)

ex-	pro-	con-
de-	co-	syn-
retro-	semi-	meta-

(2) Which prefixes can be combined with the four words to give new meanings?

Suffixes should be the source of added delight for the active writer. "-Ish" and "-ness" attach handily to the given word–pottish, potness; testish, testness; cupelish, cupelness–the applicability or suitability of such terms aside. Government, church, and legal representatives engender great activity in suffixation in writings. Unfortunately, the belief is held by many that such use is appropriate, formal, and authoritatively acceptable. Ergo, *-atic*, *-ation*, *-ative*, *-ific*, *-istic*, *-ate*, and *-ism* render the content of some documents the full measure of complexity, if not nonsense. Yet, despite that trend (or because of it), we must all be aware of the implications of such usages. Where, for example, would the *Emancipation Proclamation* be were it called the "Freeing Announcement"?

Dosage 28: Suffolksing . . .

Define and combine the following suffixes.

-able	-er/-or	-ic	-ific
-al	-ile	-ical	-istic
-ly	-ory	-wise	-ize

Note that some combinations unnecessarily lengthen a word when a simpler form would do (epidemiologic instead of epidemiological). What is the difference between "historic" and "historical," "classic" and "classical," "economic" and "economical"?

SIMPLE SIGNS

In our seeming traverse back to the grunt and gesture, we are obliged to consider the trod, the sod, and the word signs in our reckoning. In the early scrolls of written language, no spaces were given between letters, and words were not easily discriminable. Hence, YOURMIGHTYWITMAYWELLHAVEBEENTESTEDTOPERCEIVETHEPOINT.

Perception is critical to understanding, as this doodle, found in a coed's notebook, gives witness:

Fig. 10-1. Cover the bottom half and discern the universal principle.

At first splash, the figure seems foreign, but when you hold a pencil across the bottom half of the doodle, you can see the message, give meaning to the doodle, and perhaps even become communicated with by the selfsame coed.

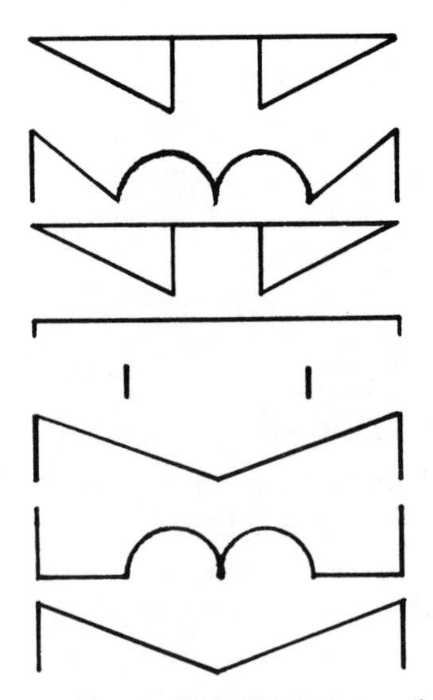

Fig. 10-2. Turn the page ninety degrees clockwise, and cover the bottom half of the figure. Do not call before 6:00 P.M.

Hyphens

Inseparability is tenacious, and the atavistic grunt-move seems now to disdain some of those devices that cued the reader about specific word meanings. Whither has gone the hyphen in cooperate, in pre-exist, in firsthand? and whither the space in percent? Consider the possible difficulties for the reader in these phrases: Is a suit-minded person a lawyer or a tailor? How can you interpret these combinations?

suitseeker bidder

(bridge-player or politician seeking to recoup insurance money by bidding on a surefire loss? or a suit seeking a bidder?)

wedding suit

(breach of promise or promise of breeches? or a divorce?)

Some languages have word endings to cue the reader about a word's meaning in a statement. For instance, had the Latin-speaking Romans known then what we know now they would have had no difficulty in forming an unmistakable expression for what we term "slow virus infections." Many readers might believe that these infections result from "slow viruses." Nix. The reality is that these infections, caused by viruses, do not produce any symptoms for many months or even years. Slowness, then, pertains to the infection, not to the virus. Yet, scientists, even virologists, continue to refer to "slow viruses." Could the difficulty be resolved by Gajdusek's recommendation that the agents be called "unconventional viruses"?

Were we all better schooled in the use of the hyphen and other devices, much of the ambiguity in language could be prevented. Small bowel lesion and small-bowel lesion differ in that we know the size (comparative) of the first and the more specific location of the second but not vice versa.

In general, hyphens are used:

(1) to join two or more words as a single modifier preceding a noun (small-bowel lesion, a three-year-old girl);
(2) to indicate a particular meaning of a prefix and thereby avoid ambiguity (re-create);
(3) to avoid awkward spellings (micro-organism);

(4) to join the numerator and denominator of fractions used as adjectives (one-half) and numerals with units of measure used as adjectives (3-ml tube);
(5) to form compound numbers from twenty-one to ninety-nine; and
(6) to coin new words by joining two or more words (noun-sense).

A dozen or more refinements* apply to the above general rules. For example, if the words function as a unit modifier but they modify the subject of the sentence and occur after the verb, the hyphen is not used. (The girl is three years old.) If one of the two words functioning as a unit modifier is an adverb ending in *-ly*, do not hyphenate. Fractions used as nouns need not be hyphenated (three fourths of the bottle). If a scientific term, such as the name of a chemical or disease, does not have a hyphen in its original form, do not use one even if the term occurs as a unit modifier (methyl bromide solution).

Apostrophes

Granted, the hyphen business seems to rise and fall with the economy; at times hyphens flood the market and at other times they are scarce. Their usage seems to correlate with the consumer's perceived need for them.

But the *apostrophe*! Oh, Apostrophe! You are a commodity that is a staple of everyone's possessions and a quick cut to their meanings. Every child uses apostrophes. How can it be that grant applications for hundreds of thousands of dollars, written by men of learning, are disjointed by the misuse of you, Apostrophe?

Let us address ourselves to the most frequent abuse of all—the confusion of "its" and "it's." (Yes, believe it or not.) No doubt the confusion arises from the fact that to show possession by a *noun* one adds *-'s* or *-s'*. Indeed. However, "it" is a *pro*noun, not a noun, and pronouns show possession by changing their forms, not by the apostrophe.

*The most thorough discussion, examples, and lists are provided in the U.S. Government Printing Office *Style Manual* (January 1973).

You remember these possessive pronouns:

my	our
your	your (y'alls)
its	their

Contractions, short cuts to word usage, usually verbs, indicate the omission of a letter or two, which allows the speaker to slur two words together as one: it's, didn't, can't, y'all.

Here are a few rules to forever defuse your confusion about "its" and "it's."

- Do not use contractions in your formal writing. Contractions are primarily for spoken language and for a casual, conversational writing. Standard written English does not include contractions.
- Can you say "it is" and mean what your "it's" means? If so, use "it is" or, better yet, rewrite the sentence to omit such a meaningless expression.
- Can you say "'tis" and mean what your "it's" means? If so, apply the above recommendation for "it is."

A third use of the apostrophe that often enough puzzles the science writer is to indicate a plural coined of letters, figures, and symbols. English grammars and the government style manual still recommend using the *-'s*, and so do we.

1950's	SCOR's
ABC's	2 by 4's

However, inasmuch as some journals are seeking to trim away every imaginable space in a line to cut costs, 'tis conceivable that such an apostrophe could be edited out (alas) of your printed article.

SIMPLE COMPLEXITIES

Enter now the delightful realm of combining words to express thought-clusters. A Chinese perspective on the horse and rider is that the two are in fact three: horse, rider, and the two together. That perspective holds for word clusters called phrases.

Adjectival Phrases

The simplest legitimate word cluster is the phrase that describes a noun-sense:

a cell
a blood cell
a red blood cell
an abnormal red blood cell

The grouping expands to the degree necessary to describe the noun-sense for the reader. Once identified, the "cell" can thereafter simply be referred to as such.

Whereas you may appropriately present a series of adjectives to describe a noun, beware of using nouns as adjectives in simple phrases, which may lead eventually to undisciplined pilings of nouns forced to function as adjectives, "stacked nouns."

You object, "So what, as long as the six-word phrase is understood?" Aha, but you never know if others derive the same understanding from those words. Even if understood, the phrase in its awkwardness can slow your reader down, annoy him, and reduce the effectiveness of your message.

A noun, in typical English usage, is usually followed by a verb or a preposition, or it occurs at the end of a phrase. Hence, when nouns are stacked, the reader assumes at each noun that the noun-sense is complete, only to find at the next noun that he has been tricked and must go back and reinterpret the preceding noun as an adjective. Bon, the reader says, after the second noun, I can now continue with the verb–but, again, at noun three, he finds that same sly trickery. With tried patience, the reader again converts nouns to adjectives, repeating the procedure a fourth time until he completes your subject-concept. However, even though the subject is now complete, the reader may still ask, What does that mean? ("Mean" is used contextually here as a verb. Were the sentence lengthened and adjectives stacked, it could become an adjective: What does that mean SOB heir-to-an-anomalous-wit author intend to communicate by that garble? Of course, he still must determine here whether "mean" means "malicious," "petty," or "stingy." Probably neither. The author is not stingy with words, and since he is not conscious of what he is doing to the reader, he is not willfully evil.)

In trimming away adjective-noun joints, ask whether real adjectives exist that can substitute for the nouns that are being forced into an adjectival position. If not, try reorganizing, maybe simplifying, the sentence.

Have some fun with stacked nouns in Dosage 29.

Dosage 29: Obfuscating . . . (come again?)

To suitably obfuscate your writing for politics or law, mechanically select terms from the list below. We prepared this easy checklist of random-selection terms with embellishment of affixes for your ambient parameter modality utilization. Words may be selected in any order from any of the columns. Start by choosing an A, B, C, D order. Devise at least three compound terms:

________ ________ ________ ________

________ ________ ________ ________

________ ________ ________ ________

Combine Nouns to Suit Obfuscation Need*

	A	B	C	D
(1)	finite	system	di-triad	capability
(2)	ambient	method	effect	analysis
(3)	response	pile	affect	evaluation
(4)	proreactive	document	modality	pursuit
(5)	distributive	detail	temporality	development
(6)	significant	parameter	dysfunction	presentation
(7)	primary	methodology	stimulus	utilization

---

*Add prefixes *non-*, *mono-*, *xeno-*, *ab-*, *co-*, *pre-*, *re-*, and *inter-* and suffixes *-ate*, *-ize*, *-atory*, *-ologym*, or *-atrology* to suit sound-comfort need ad lib.

Prepositional Phrases

The prepositional phrase is so called because it begins with an indicator word prepositioned before a noun-sense word.

at the juncture
with a scalpel
in the beginning
after mitosis

Note that the preposition answers to such questions as where? when? and how? You are reminded of adverbs, of course, which specify something about verbs, adjectives, and other adverbs. Likewise, prepositional phrases elaborate on verbs, adjectives, and adverbs, telling us *where* the surgeon *put* the clamp (at the juncture), *how* he *cut* into the skin (with a scalpel), *when* anesthesia *was administered* (in the beginning), and so on. As modifiers, prepositional phrases should be placed as close as unawkwardly possible to the words they modify.

Scientists who want to be specific in describing a process tend to overload sentences with these phrases. Care should be taken that all phrases are essential to the meaning of the sentence and are not redundant to the concept they are describing. For example, if the solution turns "blue," it need not turn blue "in color."

A confusing pileup is the prepositional phrase that in turn is modified by a prepositional phrase, that in turn is modified by a prepositional phrase, that in turn . . .

> Dr. Schmouser believed that white rats *in early life from east-coast labs with black spots in the center of their noses* were appropriate *for the experiment.*

Pruning such a sentence removes the extranea, provides a more logical form, and gives the reader a briefer, more understandable message. To improve Dr. Schmouser's prepositional phrases, put them in their most logical order; then, convert them to simpler forms if possible—for example, try replacing them with adjectives, adverbs, and possessive forms. Briefer, equivalent words for the phrases are "young," "infant," and "newborn" for "in early life," and "black nose spots" for "black spots in the centers of their noses."

Should all the prepositional elements be retained in the sentence or are some meanings so obvious from the context as to be superfluous? Which elements can be discarded? "in the centers"? "for the experiment"? Here are two editors' revisions of the example:

(1) Dr. Schmouser believed that newborn white rats with black nose spots and bred in eastcoast labs were appropriate.
(2) Dr. Schmouser believed that white rats were appropriate if they were newborn, had black nose spots, and were bred in eastcoast labs.

Dosage 30: Reject the preposition . . .

Prune these sentences by reducing the number of prepositional phrases (italicized).

(1) In the absence *of an increase in the amount of insulin*, an increase *in the level of catecholamine* would increase the concentration *of cyclic AMP with the later enhancement of glycogenolysis.*
(2) The results *of the study on lipids* were returned *to the students by campus mail in their freshman* year and again *in the senior year along with a questionnaire for getting information on the amount of weight change*, if any.
(3) A growing awareness *of the importance of digestive diseases on the part of the government* has resulted *in an increase in research appropriations in recent years.*

Verb Phrases

Not to be outdone, active words can also cluster:

to write	will be written
was written	could have been written
has been written	will have been rejected

The proper verb cluster will depend on the context. In science writing, the options usually will be limited to observations that were made in the past, or to conclusions that are in the present, or to studies that will be done in the future.

Chronology notwithstanding, science writers seem especially inclined to use a verb form that, by the mere addition of *-ing* to the verb, yields flexibility of meaning and timeless possibilities for clustering. One use of such a verb-sense word, called a "verbal," is in a noun-sense, called a "gerund."

Writing helps.
Helping hurts.
Hurting hinders.

Because the *-ing* form of a verb is called the present participle and the *-ed* form of a verb is called the past participle, word clusters using *-ing* and *-ed* verb forms are sometimes called participial phrases. Participial phrases are verbals used in an adjective-sense.

Having published the article, he awaited the reactions.
Writing furiously, she finished the editorial.
Written science articles communicate important information.
Clustered words interrelate.

Because the participial phrase functions as an adjective, the noun or pronoun it modifies must be expressed in the sentence, that is, something or someone whose action or state of being is described. In addition, because participial phrases are verbals, they retain their verb-sense characteristics of expressing the verb idea, taking any kind of complement, and being modified by adverbs. But do not let all that activity confuse you; there must nevertheless be the something or someone in the sentence to modify—and often that something or someone is the subject of the main clause of the sentence. Including and distinguishing that modified entity in the sentence is the critical factor in preventing misuse of the participial phrase as a "dangling modifier."

Participial phrases have a way of hanging around in awkward places in sentences, and locating them with an expressed subject may require completely reworking the sentence:

Rising to the occasion, his chair slipped from under him.

In this sentence, the rising was seemingly done by the chair. The reader is left to figure out that what was really meant is that someone's chair slipped as he rose.

It does not suffice for the writer to know the doer. Not only will the meaning be ludicrous without a subject expressed, but also the reader may not be able to correctly fill in the blank for you. So, avoid risk and save face:

> Rising to the occasion, the toastmaster slid from his chair onto the floor.

When such misuse is ludicrous, it usually is obvious; unfortunately, the misuse is often subtle:

Wrong: Sweating profusely, his nurse wiped his brow.
Right: Sweating profusely, he had the nurse wipe his brow.
Better: The nurse wiped his profusely sweating brow.

Wrong: Closing the wound, the surgeon's stare was fixed on the chief resident.
Right: Closing the wound, the surgeon fixed his stare on the chief resident.
Better: The surgeon closing the wound fixed his stare on the chief resident.
Better Yet: The surgeon fixed his stare on the wound he was closing (!).

But note the subtle misuse in this example:

Wrong: *Judging by current standards*, the procedure was well done.
(The procedure is judging in this sentence!)
Right: *Judged by present standards . . .*
(The procedure is judged, which is sensible.)
Better: The procedure was well done according to (by) present standards.

Wrong: *Collected in a bottle*, we had the pills sent forth.
Right: *Collected in a bottle*, the pills . . .
Better: The pills collected in a bottle were sent forth.

Dosage 31: Dangling at the end of a rope, I . . .

Underline the dangling participial phrases in the examples. Then rewrite each sentence correctly in your notebook.

(1) After resting fifteen minutes in the sitting position, a baseline arterial sample was drawn.
(2) All hemorrhoids should be carefully inspected before making any incisions.
(3) Using a balance beam, weight was measured for ambulatory patients dressed in light clothing wearing no shoes.
(4) On regular chow and using an interrupted schedule, practically no sterols were detected in the plamsa of the rabbits. However, feeding a mixture of plant sterols at 2 percent level of ten weeks, measurable quantities of campesterol were present in the plasma of all rabbits.
(5) The pylorus was ligated with triple 0 silk suture using metofane for anesthesia.
(6) Dying prematurely, the patient's family can undergo extreme distress.

PART III. LANGUAGE STRUCTURE AND MEANING: PASSAGES AND MESSAGES

11. Formulas of Language

PATTERN

All things have structure—whether a chair, a painting, a mucopolysaccharide, an aardvark, our solar system, a computer program, a formula, a word, a sentence, or language. The structure is the arrangement of parts or elements that give the item an identifiable shape—the formal part of its essence—identifiable either by appearance or by function, or both.

Man forms structures out of wood and brick, glass and steel, words and words. We put things together and thereby establish a spatial relationship between the objects or the words. That spatial relationship forms a pattern, and if we position the elements carefully enough, that is, if we *juxtapose* them, we can again form new patterns and give new meanings to the relationships. Consider, for example, the kaleidoscope you had as a child. Within the base of the kaleidoscope are a few pieces of colored glass that reflect off mirrors. Yet, each twist of the base repositions the pieces and changes the reflections, so that a new pattern forms.

The kaleidoscope, in showing an infinite number of abstract configurations, illustrates how a few bits and pieces can produce such patterns. The few words used to express one thought can also recombine to form new patterns of meaning. Imagine the potential combinations of hundreds of thousands of words in the English language.

Pattern is form that can be repeatedly applied; it underlies replication of logic, of research method, of experimental results, and of communication. For example, to repeat an experiment, you must know what steps to follow in what order. In mathematics, theorem proofs are repeatable and follow specific and systematic, if not absolute, steps toward an inevitable conclusion. Each step involves moving information from a preceding step to a succeeding one, i.e., tracing the logic of the concept. For experiments to proceed logically step by step, the method must be described accurately and concisely,

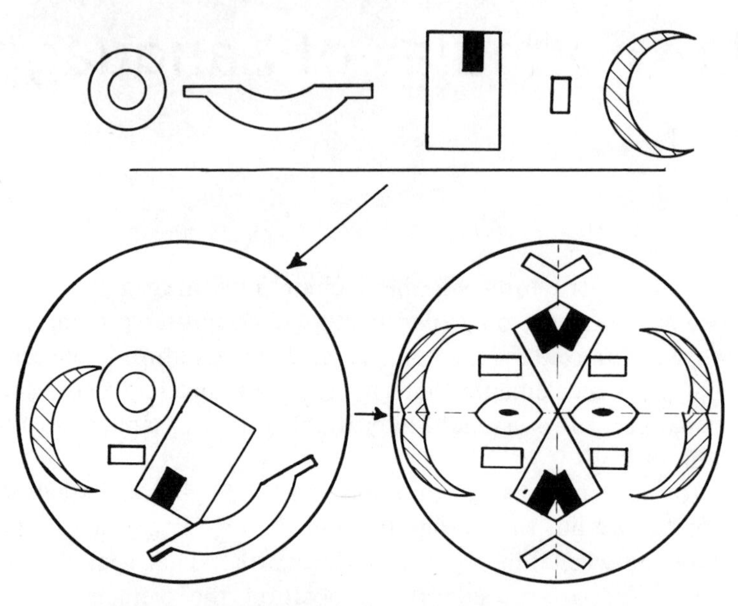

Fig. 11-1. Patterning is important in the kaleidoscope image.

which means that each step in the description of the method—each sentence, paragraph, and section—must be concise and accurate.

Likewise, in language, if words, word groups, and sentences are to have consistent meanings for the reader, they must group together in a consistent way, in patterns that are recognizable, so that the idea is consistently repeatable, can be built on, and can lead to the next thought-step.

Example:	tube	test tube	clamp test tube
	clamp	tube clamp	test tube clamp
	test	clamp test	tube clamp test

In language, the patterns have existed for centuries. But for most people, because language is second nature, the patterns are not easy to identify. We tend to rush through words to get a message and, focusing on the meaning of that message, we overlook or simply are unaware of the pattern the words are forming to deliver that message. We do not consider how the words combine to transmit a specific idea. Even if you try to recognize the pattern, chances are that your

Fig. 11-2. Vase or faces? Background and foreground switch.

mind will become distracted or confused by the individual word meanings, and you will not be able to identify the formal order of the word groups. The old saw about not seeing the forest for the trees might be re-combined to point out that neither does one see the trees for the forest.

Fig. 11-3. Perception involves the parts in the whole, and the whole of the parts. Can you see the forest for the hills and rock?

Or, if you are a city folk, you might be more comfortable with a laboratory illustration. In looking at a slide of cells, you can see numerous, individual cells, or you can see a configuration formed by a group of cells. For some diseases, the number or type of cells might be the significant information; for others, the patterns.

Yeats inquired, "How can we know the dancer from the dance?" How can we know the image from the paint? How can we know the disease from the signs and symptoms? Can we know the meaning of a word without its sound or sound symbol? Can we communicate an idea in words without a structure for the sounds?

Sometimes we do not even know the meaning of an idea that we have in our minds until we give it a form—by putting it in writing or by speaking it—and in the very act of giving it form, the meaning becomes clear to us.

"Juxtaposition" comes from "juxtapose" meaning to put side by side or together. Juxtaposing words, ideas, illustrations, and forms side by side is a technique artists and poets have used for years. They have understood that not only new thoughts, but also communication is achieved by such juxtaposition. Deliberate combinations, or juxtapositions, of words can range from the unique, as in poetry, to formalized patterns, as in traffic signs: "stop" is conveyed by a hexagon, "yield" by a triangle. Internationally adopted signs use pictures or symbols instead of words—for example, a slash means "no."

In science articles, purpose, methods of investigation, results, and discussion are juxtaposed in a pattern that leads the reader's mind to

Fig. 11-4. No smoking in or near this book.

derive order out of what could be chaos. Formalized structures can relieve the reader of certain burdens of thought and allow him to get on with the new subject matter at eye. In other words, if the first step, whose symbol is universally recognized, has been established, the reader can hop quickly to the next step and go on with the idea.

Dosage 32: Symbolically speaking . . .

What are the ideas symbolized by these signs?

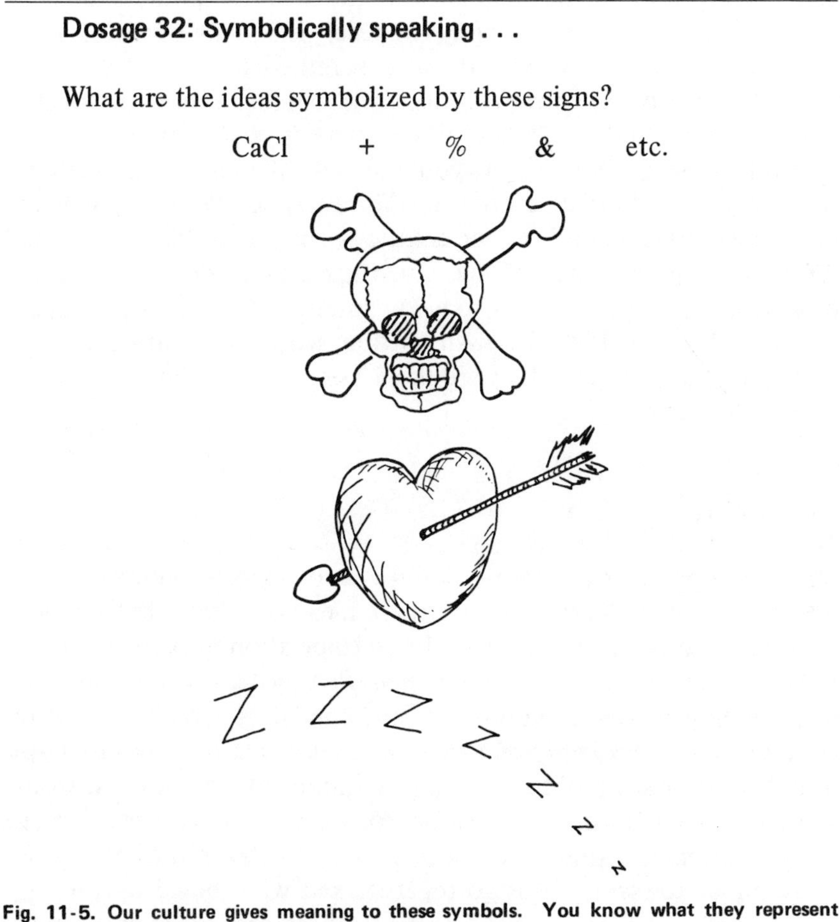

Fig. 11-5. Our culture gives meaning to these symbols. You know what they represent without having to be told.

Today, the path to novelty, creativity, and intellectual excitement has become so widespread—look at magazines, movies, music—that one becomes easily bored with mere linear presentations. The mind craves depth and variety in learning situations.

This imposing overabundance of distractions and stimuli poses a problem for scientists. Science publications are still linear in the extreme. And the problem is that instead of capitalizing on that linearity by striving for utter simplicity (i.e., unity of thought and straightforward expression), by not appreciating the linearity, authors unnecessarily complicate their writings. The result is a piling of words into a puzzle that diverts the smooth progression of thought. In a linear form, true clarity of thought proceeds from ideas expressed directly, as simply as possible.

In a laboratory experiment, you add only the chemicals necessary to produce the desired reaction in the test tube. Perhaps you could add many other conditions or solutions to the method that would not change the outcome of the experiment, but what for? The simplest technique, the unsuperfluous technique, best meets the experimental objective. Likewise with writing, why add word conditions, patterns, and structures that are unnecessary and that delay communication of meaning?

GRAMMAR

Out of the preceding regimen of forms and patterns, you should be able to distill at least one important message: that what happens among words or elements linked by juxtaposition is as important as what happens in the words individually. A system, steeped in tradition, already exists to describe those happenings—grammar. Grammar, which you probably remember as a set of tedious rules that you would rather not be bothered with, is simply the way we put words together to let another person see or hear what we want him to know. Grammar is not really a mere set of rules; rather, it is a description of the way words go together and why, based on the way most people speak and write. Grammar, therefore, is not prescriptive but descriptive—and "correctness" varies from situation to situation, from casual communication to formal, written to spoken, and language to language.

Three grammatical elements give language meaning:

- Syntax, or word order (formulas)
- Inflection, or word changes (morphology)
- Punctuation, or signs of meaning

Scientists can usually relate easily to mathematical and chemical formulas and promptly see the structure beneath. Consider the following examples.

(1) $20 - 4(2) = 12$
(2) $x + 2(y - z) = k^2$
(3) $HCL + NaOH = HOH + Na^+Cl^-$
(4) $Ba + 2Na + S = BaNaNaS$
(5)

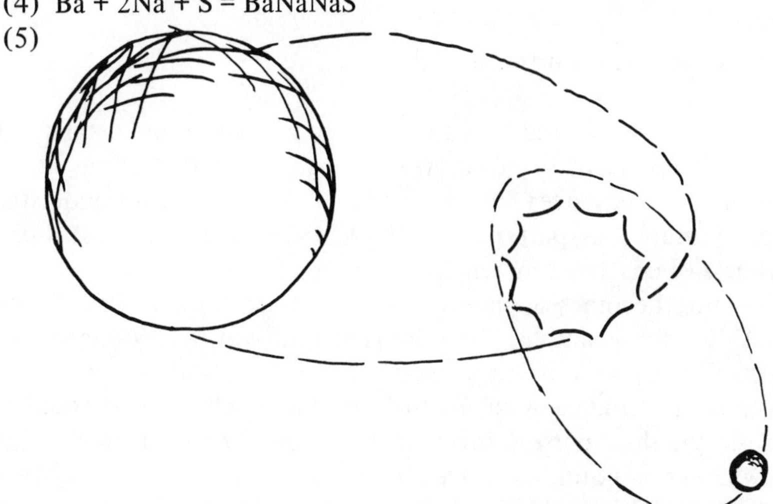

Fig. 11-6. Is this your idea of an atom?

The numerals and chemical symbols represent basic structures undergoing movement that brings them to a different state or concept, that is, the idea of the total "12" evolves from the individual number units and steps of the equation, and the chemical compound evolves from the interaction of the positive and negative charges of distinct elements. Signs show the direction and type of movement, and each sign is a convention agreed on, defined, and used by everyone wanting to communicate through mathematics and chemistry.

Language is no different. A sentence is a formula that moves a thought from an original state and partial concept to a new state, or a complete concept. A sentence has conventional structures and signs that tell the reader the direction of that movement and how the various units combine. The sum of those structures and signs is different from its units. You have simply to learn to recognize the individual units and the order in which they most logically combine with other units to perform certain functions. Those functions will partially or wholly be indicated by the order or pattern of the units, and partially by conventional signs, that is, punctuation. (What, ho! What hoe?)

Syntax: Sentence Formulas

By now, syntax should be a concept you can easily understand. Syntax is the pattern, or structure, of word order that forms phrases, clauses, and sentences. We believe that the writer who understands those patterns, as patterns, and their underlying logic, will find the business of constructing sentences an easy-flowing activity.

Meaning becomes apparent in a syntactical context, in a formula. Consider, for example, the following two words: "communication means." "Communication" is a noun or it could be an adjective; "means" is either a noun in the singular or plural or it could be a verb in the third person singular form. How the words function, that is, whether as noun, adjective, or verb, depends on how they are juxtaposed to other words in a sentence. In the context of the sentence, "Communication means exchange of ideas," the relationship of the two words becomes apparent: "Communication" is a noun (the subject) and "means" is a verb (the predicate). If, however, a verb follows both words, "Communication means are newspapers, television, radio, satellite, and belchaseltzer," then the two words together, contextually, become a subject unit, and "communication" functions not as a noun, but as an adjective, that is, as a specifier of "means"–it tells what kind of means. "Means" functions as a noun. (If one is to respect the form of the words and use them in the strict and precise way as a subject unit, both "communication" and "means" should remain nouns: "means of communication.")

Dosage 33: Scribe words

Can you see multiple possibilities for the following words?

ground hogs
house flies
government steps
tool works

Cell Blocks

To continue, we will identify the most usual word units and the most usual patterns of units; then, we can see how variations of those patterns change the movement of the sentence, and why you might want to change the movement.

Conceptual movements from clause to clause within a sentence are similar to the movements from sentence to sentence within a paragraph. What is true about sentence units and the way they communicate ideas can be applied also to paragraphs. Thus, you get double your investment in applying your mind to the following section and dosages.

In traditional terms, sentences are *simple* (you might want to take issue with tradition on that), *compound*, *complex* (Aha!) and *compound-complex* (mon dieu–voo doo!). We will not discard those terms, because they have clear and specific meanings, but neither will we rely solely on them. Rather, we will identify a universal principle that underlies those terms, universal in that it applies not only to sentences and language, but also to mathematics, chemistry, medicine, and other sciences and disciplines. One approach to identifying that principle is the use of diagrams.

A diagram is a formula that uses lines, spaces, points, and angles to demonstrate a mathematical proposition, to represent an object, or to show the relationship between parts. Actually, diagramming is fun and gives you a means of doing serious work while playing. We hasten to mention here that we are not necessarily talking about the

Fig. 11-7. Raillery has its place on an interdisciplinary track.

traditional method of diagramming sentences you may have loved or abhorred as a youngster in grammar school, although such diagramming is one good way to visualize the function of units within a sentence. Other visualizations more personal to your own way of perceiving the world, or just more fun, are also useful. We would like you to develop your own formula for picturing what happens in sentence units. So, for Dosage 34, grab a pencil and do some diagrammatic, formulaic thought-splashing.

Dosage 34: I diagram to say . . .

(1) You, an epidemiologist, want to develop a matrix for the variables you will analyze statistically in a study of canker on the sunny side of the street. What does your visualization tool, your diagram, look like?

After you complete the diagram, refer to p. 126 for our concept. What term is standardly used to identify the smallest consistent components of the diagram?

Now make a second diagram—this time for Legionnaire's Disease, or for some other disease in your ken.

(2) As a biologist or a pathologist, you are studying a photo-

Fig. 11-8. Here are the first few lines to get you started with your epidemiologic design.

micrograph on which are shown the smallest units of life. Draw three types of these.

(a) dendrite (a nerve-fiber ending) (b) epithelial (c) muscle

What are these tiny units called?

Now multiply and combine and recombine these units into meaningful structures, or patterns that render them identifiable as tissue or disease configurations.

(3) You are a monk in a monastery, a soldier in a fort, or a prisoner in a jail, and you have a small cubicle to lie in. Draw that cubicle.

What is that cubicle called?

Now combine multiple cubicles into visual patterns that form identifiable "housing" structures. How is a blueprint a formula?

(4) You are a bee returning from a flight out into the fields where you have collected nectar. Draw the structure you will store that nectar in.

What is the name of each tiny unit in that structure? How is a beehive a formula and a diagram? How is genetically determined behavior a formula or a pattern?

(5) You are a battery. Draw a receptacle containing two electrodes and an electrolyte.

What is the name of that receptacle? How is a battery a formula?

(6) You are a science writer writing a perfect sentence. Write that sentence.

How is each functional unit in that sentence a "cell"? We will use that word, our *cell*, as our *verbus operandi*. What larger units do these cells combine to form (cell blocks)? How do cell blocks combine to form sentences? How is a sentence a formula?

An epidemiologist, in constructing his matrix of dependent and independent variables, each of which falls into an individual cell, develops a useful diagram—a grid, a cell chart (Fig. 11-9). The biologist or pathologist studies combinations of epithelial cells that compose layers of cellular tissue (Fig. 11-10). The monk lives in a cell that is part of a monastery, a cell block that takes up a city block (Fig. 11-11); the soldier's fort is blockhouse (Fig. 11-12); the prisoner lives in a cell block (Fig. 11-13). The bee returns to his honeycombed storage cell (Fig. 11-14). The battery, a storage and power cell, generates electricity (Fig. 11-15). The writer uses brain cells but sometimes develops mental blocks (Fig. 11-16). As for the perfect sentence—that was Hemingway's haunt not ours.

INCIDENCE OF CANKER

	Street				
	MAIN		MAPLE		
Variable	Sunny	Shady	Sunny	Shady	Notes
Season					
Spring					
Summer					
Fall					
Winter					
Subjects					
Male					
Female					
Age					
0-5					
6-10					
11-15					
16-20					
21+					

Fig. 11-9. Expanded version of epidemiologic design.

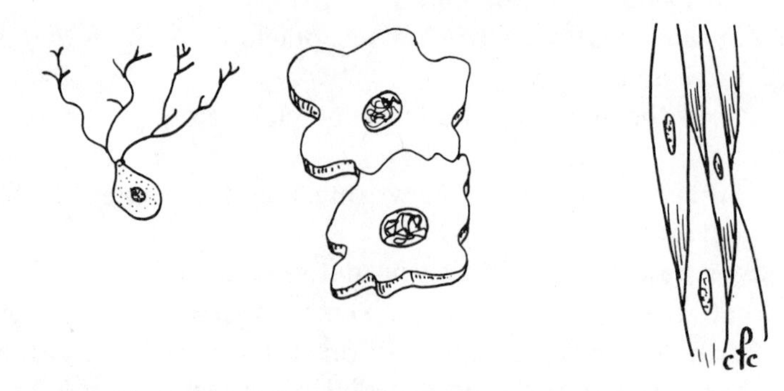

Fig. 11-10. Three body cells: nerve, squamous, muscle (sort of).

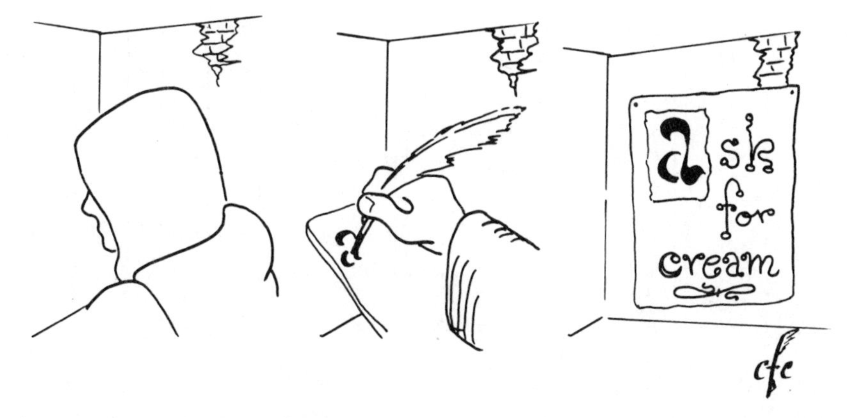

Fig. 11-11. A-monks the cloistered set, "Monk . . . writes . . . manuscript." In reverse order, the message would be "Message is written by monk."

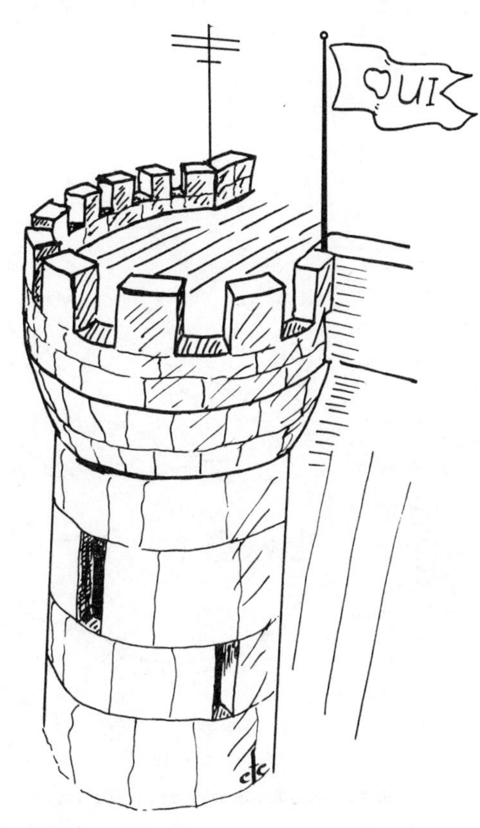

Fig. 11-12. A tower cell in a French castle for special prisoners.

Fig. 11-13. In a prison cell, "Prisoner . . . regrets . . . crime."

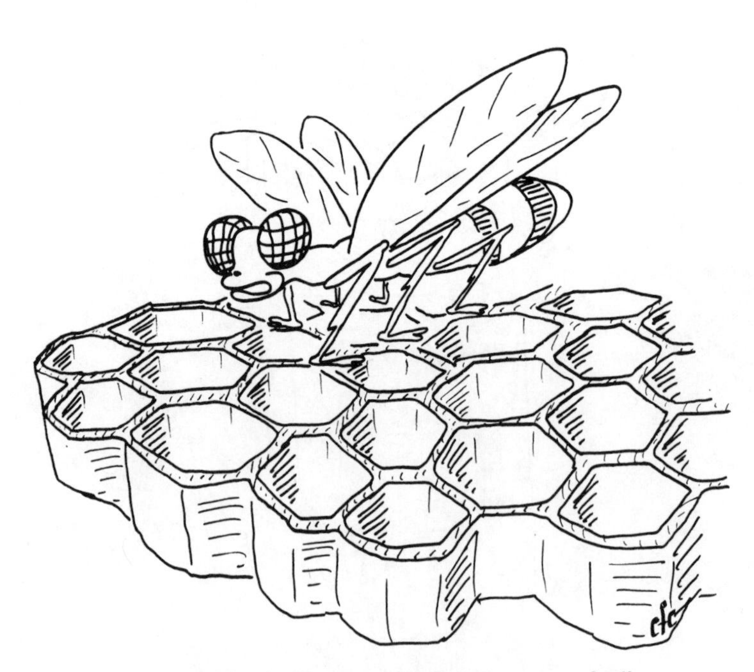

Fig. 11-14. Bee in home-comb, being a system of cells.

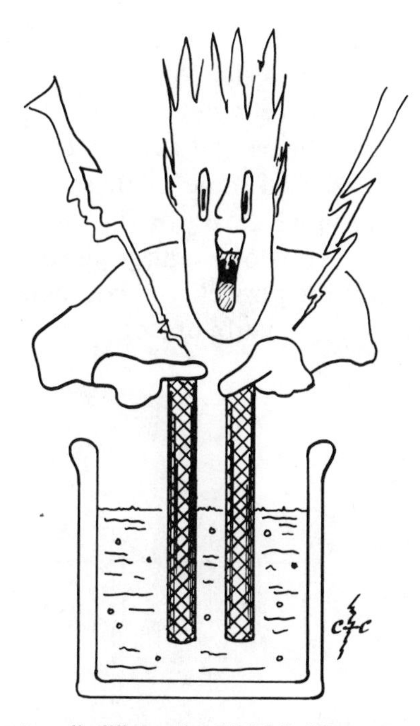

Fig. 11-15. A battery "cell" showing ? ? ? ? ? "Battery . . . shocks . . . you."

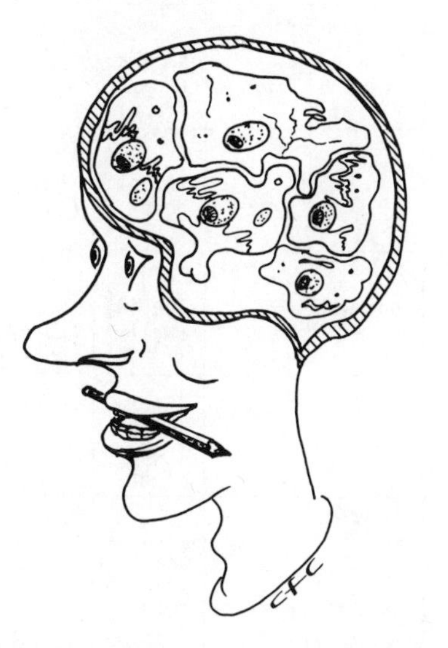

Fig. 11-16. Hemispheric cells of the new writer.

As cell blocks develop and become larger and stronger, they form blockbusters of energy—muscle cells combine into biceps, nerve cells electrify the body, statistical cells defuse mazes of confusion, and words charge the reader with ideas and excitement.

Now that we have represented some of these units with lines and spaces, we can give them more detail, or fill them with content, to see how they shape up in terms of language units. That is, for each unit or cell, let us identify the possible cell combinations that give us each such cell blocks as subject, verb, object.

1. The variable matrix

	SUBJECT	PREDICATE	OBJECT	PHRASE	MODIFIER	LINKAGE
noun (pen)	pen		pen	(with a) pen	pen (holder)	
pronoun (I, me)	I		me	(to) me	my	
verb (jump)		jump (up)		jumping (the fence)		
adjective (good)				(to be) good	good (times)	
adverb (well)		(do) well		(sleeping well)	well (care)	
preposition (with)				with (good intent)		
conjunction (although because and/but/or)						(good) but (dear)

Fig. 11-17. Variable matrix, showing how words get "blocked into" specific uses.

Dosage 35: The way I see it . . .

Develop your own diagrammatic sentence-sense drawing based on a cue from the list below most appropriate to your way of thinking.

(1) Otorhinolaryngology: How are the eyes, ears, nose, and mouth a sentence?
(2) Internal medicine: How is the alimentary canal a sentence?
(3) Cardiology: How is the circulatory system a sentence? the heart a verb?
(4) How is the lung a verb?
(5) 1853: John Snow: "Remove the pump handle."
(6) Pediatrics: Is birth the subject or verb of the life sentence?
(7) Administrator: How is a flow chart or an organizational chart making a statement?

Traditionalists and others who regarded learning with other than benign neglect will remember the type of diagram used by grammarians and teachers of English once upon a time to illustrate the structure of the sentence.

Simple sentence:

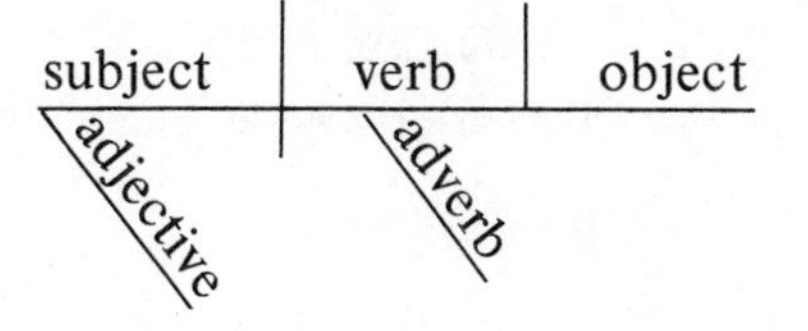

Wise apes cleverly condition researchers.

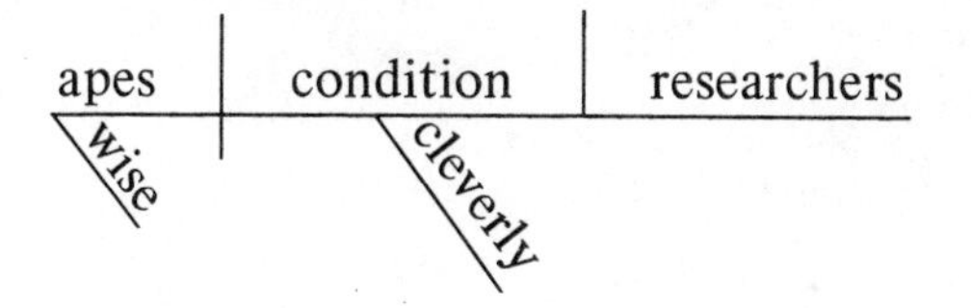

Not-so-traditional others might devise their own meaningful depuzzler, such as follows.

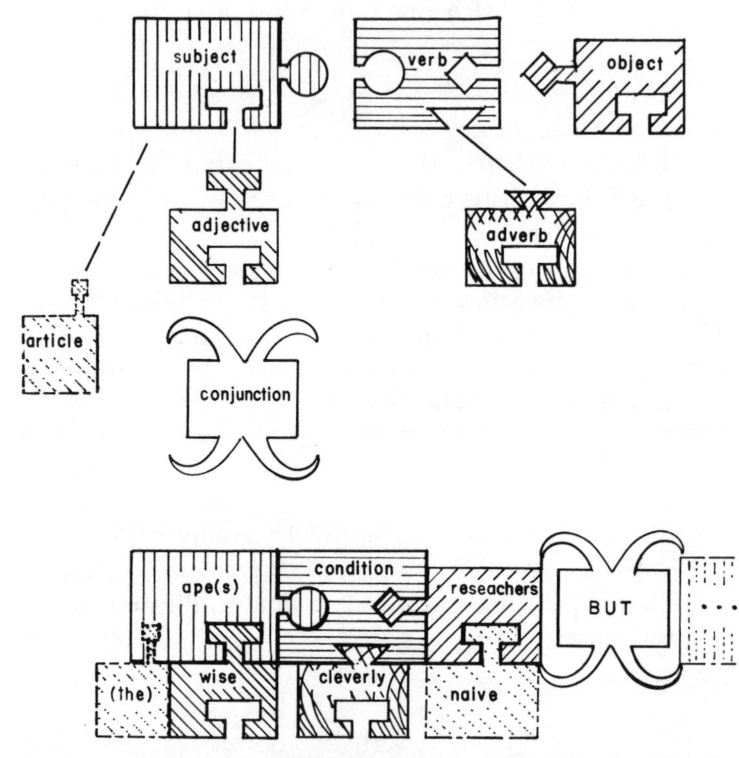

Fig. 11-18. Sentence depuzzler ("Wise apes cleverly condition naive researchers, but . . .").

12. Sentence Sense

SENTENCE MOVEMENT

However you chart it, a sentence is a series of thought cells in special sequential relation. Everything in that series depends on the verb, which is the energy cell, the action of the sentence. The verb always moves the thought cells and shows their basic relationship. The verb can move the thought in three different ways: by equating ideas, with an equation verb; by transmitting an action as a transitive verb; or by self-containing an action as an intrasitive verb.

First is the "equation" movement, achieved by *linking verbs*, or *copulas*. Consider the verb *to be:* I am, you are, it was, they will have been being, and so on. This verb actually establishes an equation and itself serves as an equals sign, that is, what follows the verb is identified with and is more or less equated with the subject.

$$B + A \overset{\text{(is)}}{=} A + B \qquad Na^{+} + Cl^{-} \xrightarrow{\text{(yields)}} NaCl$$

"The patient is a headache" is different from "The patient has a headache." "The disease is cancer" is different from "The disease simulates cancer."

A number of other verbs, besides the verb *to be*, function the same way in sentences. Also called linking verbs or copulas, they include sense verbs—feel, look, smell, sound, taste—when these verbs are followed by a word that completes the subject and its own meaning. They couple the two parts of the sentence to each other, just as the equals sign couples two sides of the equation. They complete the subject in different words, that is, with its complement. Other copulas are—continue, grow, prove, remain, turn, become, seem, appear.

The patient became uncooperative.
The patient became a headache to the physician.
The experiment proved fruitless.
The surgeon appeared troubled.
The resident turned pale.

The second kind of movement possible is for the verb to carry an action from the subject over to a receiver, an object. This transmission of action is achieved by a transitive verb, deriving from the Latin *transire* meaning to cross; hence, the verb crosses from the subject to an object, or carries the action from the subject to the object.

Ada cut the nerve.
Ada (subject) cut (transitive verb) the nerve (object).

"Nerve" is the *object* of the verb. All transitive verbs require objects: "The verb (subject) carries (transitive verb) the action (object)."

The third kind of movement is a self-containment in which the verb completes an action within itself–it is not followed by a complement for the subject, nor does it have an object that receives the action. ("The cells divided.") In other words, the action is not carried or transferred to anything, nor does it link the subject to another expressed concept as does a copula. A good way of distinguishing transitive and intransitive verbs is to repeat the verb plus "what?" Ada cut what? The nerve. Noun. Object. Transitive. If you must ask "what?" after the verb and the answer is a noun, then that noun is the object and the verb is transitive (designated *v.t.* in dictionaries). If no answer is needed, the verb is *in*transitive (designated *v.i.*). In "The tide subsides," asking "subsides what?" is ludicrous; the verb is therefore *in*transitive and needs no object. In fact, "subside" is unlike most verbs in that it is always intransitive. Most verbs can be either. Copulas are neither transitive nor intransitive.

Consider: (You) Thread the needle. (v.t., object = needle)
(You) Thread through the crowd. (v.i., "through the crowd" = a prepositional phrase modifying thread)

Motherhood is good.	(linking verb)
Motherhood is apple pie.	(linking verb)
Motherhood continues.	(intransitive)
Motherhood succeeds.	(intransitive)
Motherhood brings wonder.	(transitive)
Motherhood becomes essential.	(linking verb)

So, that was fairly simple; overall, you need know only three things about the internal movement of sentences: that they can make equations, that they can relay action onward to a receiver (transitive), and that they can be self-contained at the verb (intransitive). Within those overall movements, a few refinements can be introduced whereby your sentences develop style and achieve clarity.

SENTENCING

Some writers become confused about clauses because they are told that the simple sentence is a clause. Why not one or the other and not both? "Clause" derives from the Latin word for closure, and should convey to the reader the sense of having the thought enclosed within it. Clauses and sentences both have within their structures noun-sense words and verb-sense words that represent activity for the nouns (noun subjects and verb actions). However, the structure of some clauses is such that, although they enclose a thought, there is more to the thought.

The surgeon cuts. The surgeon (noun) cuts (verb).
But: If the surgeon cuts.

The second example has the elements of the first but also contains "if," giving the statement a sense of needing more to be complete. Actually, "if" shows the reader that the clause joins to something else; joining words are called *conjunctions.*

If the surgeon cuts, several.

The clause is now attached to something, but you can see that something is not complete. The clause needs to be attached to another clause.

If the surgeon cuts, several *things will happen.*

By now, you can appreciate that the italicized words, *things will happen*, form a clause that means something by itself. *Things will happen* is an independent statement, an independent clause, a sentence.

An incomplete sentence presented as a sentence, that is, with a capital and a period, is known as a *sentence fragment.*

Which of these is a simple sentence and which not?

Who does it?
That can be removed without difficulty.
Go.

The first example is a sentence because it ends with a question mark and makes full sense as a sentence. How would it be without the punctuation? Incomplete, dependent on a missing clause.

The second example could be either a sentence of a modifying statement, a dependent clause. The sense depends on whether you read "that" as a pronoun or an adjective. Consider:

I am bored by tissues that can be removed without difficulty.

The third example is a sentence in which the noun-subject is understood to be "you." "(You) go." The direct address is becoming increasingly popular in science writing for giving the reader specific directions or steps in a procedure: "(1) Take 1 oz of bourbon. (2) Add four drops of bitters. (3) Carefully and deliberately, spill . . . "

Remembering the name labels for the clauses is not important; knowing whether the sense is complete is important.

A sentence is somewhat like a Nekker cube. With all of the sentence elements in place, the sense may seem to represent several perspectives, or even to represent an "illusion," a joke on the reader/viewer. Even within precisely written sentences, some readers, because of experiential differences or (for want of a better word) "gists," will perceive meanings other than that blatantly expressed in the sentence form by you.

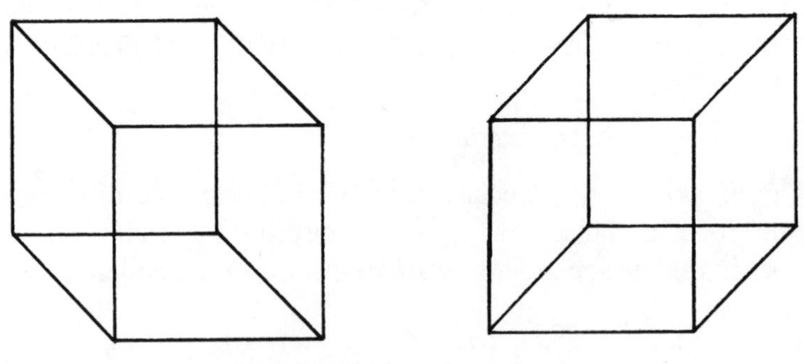

Fig. 12-1. Nekker cube, revisited.

Take another look at the Nekker cube. Can you find the Oliver spindle? Here is it, removed:

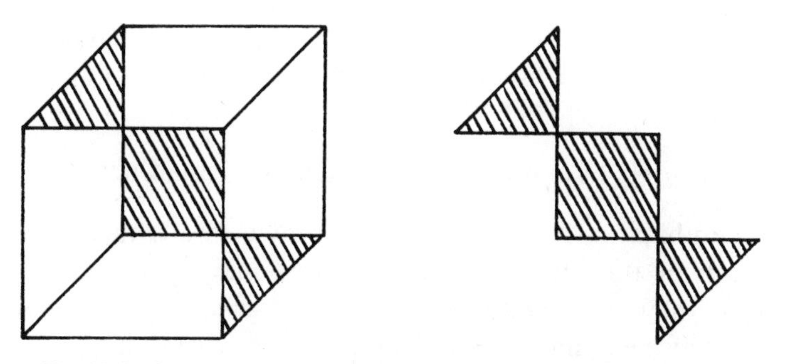

Fig. 12-2. Oliver spindle divined from within the Nekker cube illusion.

"Seeing" it rather corrupts the illusion given by the Nekker cube itself; the spindle in the figure gives the whole a "jewel with facets" appearance. Had you ever seen that before? The point of all this, since you ask, is that nothing can be too obvious. As clearly written as you may think your sentence, someone is likely to get a meaning from it other than the one you intended.

In sentencing, that flexibility and latitude in interpretation is a problem–and a help–to the writer. If you state, "Roll down the car windows and roll down the hill," you do not mean to do with the hill precisely what you did with the windows, despite your saying so. Most readers will appreciate the different interpretations in the verb "roll" and many will even see the bit of humor in the statement. Usually, however, you will want to reconsider sentence order in those expressions that imply things they should not.

The best you can do is to take care. Care is a sine qua non of good science writing, care in the selection of contents, care in the choice of words, and care in the order in which those words are given.

Simplesentencing

Whenever you find yourself groping for a start in writing, try putting your thoughts in simple sentences. A simple sentence has only one clause. Be direct; state your thoughts simply:

> We treated seventeen patients. Heavy-particle therapy had proved to be effective. Other centers use the technique.

Later, you may see that several of the separate ideas make better sense combined. For example, the last two sentences in the example might read:

> Other centers have used heavy-particle therapy effectively.

That combination is still a simple sentence and it facilitates comprehension of the idea, but combinations may also retard comprehension if they are overstuffed or not well arranged. The answer to the conscience-examining question "How shall I render it clear?" is to separate the thoughts into simple sentences and analyze the separate parts for their relationships to one another. Then, try clumping them. Sometimes the only answer is to rely on a series of short, simple sentences, to assure that the reader will not be overwhelmed.

If the subject of the sentence is very complicated, it may cry out for a simple sentence.

> malnutrition and gastroenteric studies
> mucopolysaccharide removal techniques

When in doubt about whether a sentence is simple enough, draw a box around the whole subject and underline the verb. If the subject block is too big, write two or more sentences just to describe it. After the description is set, you can then refer to it by a single-word or perhaps two-word identification, or perhaps with an abbreviation.

Here is an idea expressed awkwardly in two simple sentences that can be improved by reclumping all the information into two other simple sentences.

> Of the 626 preschool children screened during the summer months of 1974, 67 were referred for further medical care either by their private physicians or to the local hospital for physical abnormalities. (This is approximately one in ten.)

Revision:

> Sixty-seven of the 626 preschool children (about 10 percent) screened during the summer of 1974 had physical abnormalties. These children were referred either to their private physicians or to the local hospital for further medical care.

Dosage 36: Simpler sense

Note that the sentences in the following paragraph are all simple, except for the first sentence. The author has avoided using complex sentences by containing subordinate ideas within long prepositional phrases. Identify the clauses by underlining the subject and encircling the verb in each sentence. Then, break all the sentences of the paragraph into ten simple sentences. Having done so, consider how you might reconstruct the sentences into a paragraph.

> It is obvious that the naming of new ego functions contributes enormously to psychoanalytic theory. With the advent of ego psychology, psychoanalysis has been liberated from the shackles of a strictly motivational psychology. With the discovery of autonomous spheres and apparatuses, behavior can finally be understood as the operative consequence of specific ego functions. This expansion of our view of the ego has even enabled us to solve the problem of identifying the essential elements of psychoanalytic treatment, now definable as the cooperation of the analyzing function of the analyst's ego and the synthesizing function of the patient's observing ego.*

Notsosimple Sentencing

Knowing precisely what you want to say concisely, you will want to arrange the units of the sentence, and you may want to combine two or more of the overall sentence movements into one sentence, thereby compounding and complexing the sentence.

To compound or complex a sentence, you simply use more than one clause in a sentence. If you understand where the first clause is moving, you can determine the direction the second clause must take and whether that clause will function as an engine or a caboose. If the clause is independent, it will rely on its own energy to complete an idea. (Engines puff.) A dependent clause (subordinate clause) re-

*Michel Radomisli, "The paper-writing function of the ego," *Amer. J. Psychother.* 28:278–281 (1974).

lies on the independent clause for its power. (Cabooses glide while engines puff.) There could be two independent clauses, that is, two engines running on tracks side by side or linked up together with both engines running. (Engines puff and cabooses glide.)

Compounding: Pharmacists and chemists deal in compounds, and so do writers; each adds two or more things together to get some product, the compound. Whereas the chemist uses a medium to combine substances, the writer uses connectives to bind clause to clause in the sentence compound. In effect, a compound sentence combines two or more simple sentences (called main clauses or independent clauses) by a binding word called a coordinating *conjunction.*

The two unrelated thoughts in the following sentences could have been stated independently as simple sentences, each standing alone, without the conjoining word "and."

Steinmetz designed generators and Koch wrote postulates.

The conjunction is an important sign to the reader of what to expect in the relationship of the combined parts of the compound.

If the sentence is written,

Steinmetz designed generators *but* Koch wrote postulates,

the reader anticipates a difference in the two ideas, whereas in the former sentence he expected similarities.

The common coordinating conjunctions are *and*, *but*, *or*, *nor*, and *for*. Which is appropriate in these constructions?

We shall stay __ we shall go.
The rain falls __ the flowers grow.
Troubles come __ life goes on.

If the sentence is short, no punctuation is required between the clauses. In long sentences with one of the common coordinating conjunctions, use a comma.

Sometimes an adverb is used as a conjunction between two main clauses to show transition and establish a special relationship. In such sentences the clauses (one introduced by the conjunctive adverb) are separated by a semicolon.

Rome burned; therefore, Nero fiddled.

A good test for the use of a comma or semicolon is your ear. If in speaking, a pause is brief, use a comma; if it is long, use a semicolon.

Developing the Complex: Because of the need in science writing to include complicated terms, writers tend to give obeisance to complication per se. Perhaps they complicate their sentences to place the already complicated term in a seemingly familiar milieu. Whatever, complicated encounters in science can be resolved in an uncomplicated, though complex, sentence.

The complex sentence combines an independent clause with one or more *subordinate* or dependent clauses. Take a simple sentence and build a complex sentence from it, masochism notwithstanding.

> *Simple:* Peanut oil induced muscle lesions.
> *Complex:* As Finnerty reported, peanut oil induced . . .

In effect, "as" serves as a conjunction, combining the two thoughts about Finnerty and the peanut oil, with Finnerty obviously subordinate (although named first) to the peanut oil.

The subordination of one clause to another is a lot like the arithmetic function of division—the subordinate or dependent clause has in a sense been divided out, cut out, of the meaning of the independent clause. The dependent clause enhances an idea introduced in the independent clause, and in so doing relates in a special way to the independent clause. Signs tell the reader what that relationship is. For example, think of chemical elements that form bonds; as they join to one another their individual properties change. The combined elements as a new unit, as a compound, have a structure and a valence that enable the compound to relate to the next element in a particular way. Likewise, a sentence with combined word units has bonded those elements to take on a new meaning that is different from its parts, a meaning that has its own new valence or capacity for advancing thought, that is, for relating to the next sentence. The next sentence, or word-compound, can "accept" or "repel" the thought of the preceding sentence or merely stand next to it in a series. The signals that tell you what will happen between those sentences, that is, the direction the logic will take, are the subordinating conjunctions—and in grammar these are roughly analogous to the bonds or charges in a chemical valence.

Consider, for example, the message of *and.* "And" is a balancer or an adder. It connects two more-or-less equivalent sets of word clusters, for example, in a compound sentence.

He punted *and* he panted.
Man is mortal and woman is Myrtle.

Furthermore, it can tell you that what follows is a second or third element that is equal in stature or energy to the first. If, however, you see *however*, *although*, or *but*, you know that a logical qualification or semirepulsion follows. That qualification divides out either a simple unit of the preceding idea or the entire idea itself and says that something is conditional to it. The "repulsion" conjunction alerts the reader that a differing concept is being introduced to the stated idea. Does *nevertheless* alert you to an idea that will be a balanced companion idea, a qualification, or a contradiction? The subordinating conjunction introduces the dependent clause in the sentence and can show cause (because), condition (if, although), manner (as though), or time (when, after).

Dosage 37: Trying the conjunctions

From the list, select the appropriate conjunctions for the sentences. Then check off whether the ideas in the two clauses of each sentence are similar or different, as flagged by the conjunction.

nevertheless
and
but
however
therefore
although
whereas
while

	Different	Similar
We may suffer (*but*) we will not starve.	√	
Hitler stuck the world ___ Caesar whirled his stick.		
The population is increasing ___ the rate of increase is declining.		
Nero fiddled ___ Rome burned.		
A man was murdered ___ no one saw the crime.		
The world is crumbling ___ there will be crumbs for all.		

Compounding the Complex

From the compounding of simple clause + simple clause, through the complexing of simple clause + subordinate clause, we now arrive at the confounding compound + complex combination of sentence structures. A compound-complex sentence has at least three clauses—two or more independent clauses and one subordinate clause.

> Until cortisone became available, surgical hypophysectomy was impossible and treatment for acromegaly was inadequate.

Often enough, three elements unite as a chemical compound (NaOH). Likewise, often enough, three ideas are so closely integrated to a fourth idea that they must be combined in a single sentence. The key is the close integration of ideas. A compound-complex sentence is not a mere collection of ideas; rather, it is an idea unto itself that has at least three distinct facets.

Dosage 38: Identifying structures

In the following two paragraphs, analyze each sentence by underlining the subject and encircling the verb. Each subject-verb unit represents a main (independent) clause or a subordinate (dependent) clause. Identify them. How are the clauses joined in each sentence? Are the sentences simple, compound, complex, or compound-complex? The main point of this Dosage is not whether or not you can correctly name the clauses, but whether you can see the relationships among the sentence parts.

Paragraph 1:

The old woman did not change her position until he was almost into her yard; then she rose with one hand fisted on her hip. The daughter, a large girl in a short blue organdy dress, saw him all at once and jumped up and began to stamp and point and make excited speechless sounds.*

*Flannery O'Connor, "The Life You Save May Be Your Own," in *A Good Man Is Hard to Find.* New York: Image Books, 1955, p. 53. Copyright 1955 by Flannery O'Connor, reprinted with permission of publisher, Harcourt Brace Jovanovitch, Inc., and Harold Matson Co.

Paragraph 2:

I proceeded to forget Maurice, but not this DNA photograph. A potential key to the secret of life was impossible to push out of my mind. The fact that I was unable to interpret it did not bother me. It was certainly better to imagine myself becoming famous than maturing into a stifled academic who had never risked a thought. I was also encouraged by the very exciting rumor that Linus Pauling had partly solved the structure of proteins.*

Dosage 39: Sharpen your clause(s)

A clause contains a subject and a verb, functioning as a sentence or a part thereof (e.g., as a noun, an adjective, or an adverb). Clauses combine to form *compound* (two or more independent clauses) or *compound-complex* (dependent clause(s) plus independent clauses) sentences. In your notebook, combine these clauses to form meaningful sentences, ad lib.

Examples:

That writing can be fun. Nabikov knew full well
Nabikov knew full well that writing can be fun.

Dependent Clauses	Independent Clauses
that writing can be fun	the lab will be closed
was it Heroditus who said	water freezes
unless the test is completed	Nabikov knew full well
when the temperature drops below 0°C	beware of doing good without doing well
if the litmus turns red	the solution is sour
as the humidity drops	hormones become sluicey
while winter advances	air pressure rises

---

*James D. Watson, *The Double Helix.* New York: The American Library, Inc., 1969, p. 30. Reprinted with permission of Atheneum Publishers.

WHO DUNNIT? (ACTIVE/PASSIVE VOICE)

Within the movement of any sentence pattern is a special relationship between the subject and verb, known as the "whodunnit" syndrome. Occasionally in a sentence, the subject is the passive recipient of an action and the reader has no clue about who or what made the action happen.

The patient was operated on.
The patient's sphincter of Oddi was removed surgically.
The hucklebone was broken.

Who operated? Who removed(inadvertently?)? Who broke? Maybe the actor prefers to remain anonymous.

Such nonattributive expression of the verb is called the *passive voice.* The passive voice retards and even reverses the movement of a sentence. It requires the reader's thought, and even his eye, to return to the beginning of the sentence to complete the thought. Then, to continue with the next sentence, he must jump his eye forward over the lines he has just backtracked. The passive voice functions like an inhibiting agent in a chemical reaction that reverses the reaction.

Besides slowing down communication, the passive voice may muddle. There is always the chance that the writer did not specify the actor because he was concentrating too hard on the passive subject; hence, he may not know the source of the actor, and neither does the reader.

The tubes were dried in a kiln.

Note that if you must ask "by whom?" after the verb, the sentence is passive. Often the key is given in the sentence by a prepositional phrase.

The document should be obtained and studied by everyone. (passive)
Everyone should get and study the document. (active)

Some occasions are appropriate for use of the passive, others are not.

He was shot and left for dead.

Presumably, the assailant remains unidentified and the sentence form shows that. The sentence *could* be written, "An assailant shot him . . . " but that is really no better, because the "assailant" is still an unidentified shooter. Use of the passive is appropriate for other purposes.

- To deliberately avoid attribution:
  The school was considered second rate.
- To avoid identifying minor-role players:
  The equipment was sterilized [by the technician].

The passive voice should *not* be used to avoid identifying yourself as author.

Most sentences should be written in the *active voice* if at all possible. Scientists often argue that methods and results can only be reported in the passive, but the only support for the viewpoint is tradition. The active voice propels the reader's thought and eye forward, catalyzing the movement from idea to idea, and giving energy to the style.

Dosage 40: Who dunnit?

Rewrite the following sentences in the active voice.

(1) It is estimated by the American Heart Association that 23 million adults have hypertension.
(2) The diagnosis of chronic tophaceous gout was made.
(3) The data for the 312 patients are summarized in Table 1.
(4) An enlarged edematous cervical cord was found by the examining physician.
(5) Elastosis perforans serpiginosa was first described by Lutz in 1953.
(6) Of the different dietary fats, peanut oil was found to induce lesions with fibromuscular caps resembling human fibrous plaques.

PARALLELING

The logic of expectation and the expectation of logic are the bases for parallelism in writing, that is, the use of the same form for similar elements in a series. A series, more than two elements, may be made of verbs, verbals, nouns, adjectives, phrases, or clauses. The elements have a like function in the sentence and should have a like pattern, that is, all elements should be nouns or adjectives or infinitives or participles and so on. We could even represent that pattern diagrammatically. For example, let us assign rectangles as the pattern for the serial elements in this sentence

subject block verb block serial block

From the pattern, you can readily see that the three serial elements have the same form and therefore are equivalent in their function in the sentence.

If we substitute words, the three words must be the same type.
Example:

> The programs analyze and correlate the many variables related to risk factors of hyperlipidemia, hypertension, and obesity.

Diagrammatically, that sentence might look like this:

The programs analyze and correlate the many variables related to risk factors of hyperlipidemia, hypertension, and obesity.

Suppose now that the word elements in the series were not equivalent and a participle replaces one of the nouns:

The programs analyze and correlate the many variables related to risk factors of hyperlipidemia, hypertension, and being obese.

Diagrammatically, the reader would have to fit a round peg ("being obese") into a rectangle (noun-element needed). The reader anticipates a similar shape of idea, and his thought pattern is disturbed by the change of pattern in the series.

Parallelism, the expectation that consistent parts of a thought will be completed in a consistent pattern, applies not only to words in a series, but also to phrases and clauses.

Can you make the series in the following sentence consistent? (Note that the sentence has other problems too.)

> Some of our most interesting problems encountered were one child with a one-half-inch difference in leg length, one with fairly definite rickets, and one had a highly suspicious fractured elbow.

The elements of the series begin with a noun + prepositional phrase ("one child with a one-half-inch difference in leg length"), but the third element is a full clause. We would revise this sentence first by showing that physical signs are the problems, not the children themselves, and then by establishing a series of noun phrases:

> The most interesting problems encountered among the children were a case of fairly definite rickets, a difference in leg length of one-half inch, and a suspected fractured elbow.

Be careful to consistently separate all the elements of a series from one another by consistent punctuation. Thus, although such was not the style two decades ago, all contemporary grammars rule that the comma is used after the element that is followed by "and." Without that comma (or semicolon, if appropriate), the sense of the sentence can become confused, especially with clauses and long phrases that occur mid-sentence. "And" adds an element; the comma separates elements. Without the comma, the two words joined by "and" are read as a single unit.

Consider this example:

> If all the genetic, environmental and socioeconomic factors were taken into consideration, in order to prevent just one cancer–cancer of the breast–a woman would have to be given the ludicrous advice to change her nationality a generation ago, marry an unskilled worker and have six children, two right breasts and an oophorectomy before age forty.

A series of infinitives is introduced by the word "to"–change,

Fig. 12-3. *Quelle* shape!

marry, have. The reader reads "marry an unskilled worker and have six children," as a compound infinitive structure and as item two of the series. He therefore continues reading the next words, anticipating an infinitive phrase, and is surprised by a second series with "have six children" doubling in both series. Proper punctuation would prevent the reader's confusion and spare him the time of rereading the sentence.

A revision might also take other elements of the sentence into consideration:

> If all genetic, environmental, and socioeconomic factors were considered, to prevent just one kind of cancer–breast cancer–a woman would have to change her nationality a generation ago; marry an unskilled worker; and have six children, two right breasts, and an oophorectomy before age forty.

Note that the revision adds a comma after "environmental" in the adjective series. Without it, you should read "environmental and socioeconomic" as a unit modifier and you would anticipate a third adjective before "factors"–for example:

> If all genetic, environmental and socioeconomic, and statistical factors were considered . . .

Here is another example:

> Resource centers are needed to accumulate, evaluate and disseminate training materials and techniques in aging and mental health.

To avoid reading "evaluate and disseminate training materials" as a single unit of a series with the anticipation of another infinitive, put a comma after "evaluate":

> . . . to accumulate, evaluate, and disseminate training materials and techniques . . .

In the following example, a different type of confusion results:

> The physician can be led into the subject by immediately identifying his own mind, attitudes and emotions.

"Attitudes and emotions" should properly be read here as appositives of "mind," that is, as an elaboration of what is meant by "mind."

If the three items are intended as separate elements of a series, a comma is needed after "attitudes":

. . . mind, attitudes, and emotions.

Sometimes, elements are put in serial form when they really do not belong in that form because they are not and should not be equivalent in form. For instance, the reader expects that the verbs in a series will have the same tense; otherwise they should not appear to be arranged in a series.

Wrong: The solution was heated, stirred, and will be ready when chloride salt is added.

Right: The solution was heated and stirred; it will be ready when chloride salt is added.

And be careful when you drop the auxiliary verb in a verb series that all subsequent verbs would normally use that auxiliary and that all the verbs have the same subject. If a second subject is introduced, its entire verb must be stated, even if the auxiliary is the same as in the first verb.

Wrong: Cholesterol was measured by Pearson's method, uronic acid estimated by a modified carbazole procedure of Bitter and Muir, and calcium measured by Gorsuch and Posner's method.

Right: Cholesterol was measured by Pearson's method, uronic acid was estimated by modified carbazole procedure of Bitter and Muir, and calcium was measured by Gorsuch and Posner's method.

Special attention should be paid to the modifiers of elements in a series. Does the initial adjective, for example, modify one, two, or all three nouns in a series?

He added some alcohol, saline, and just a pinch of arsenic.

"Some" seems to modify alcohol and saline, but when the reader sees that "just a pinch" modifies the arsenic, he is uncertain about it. The writer in such instances should be careful to put down what he wants to report.

He added some alcohol, a little saline, and just a pinch of arsenic.

Sentences and elements in lists, outlines, and tables should also have parallel structure. The reader, shown a list of items in which each but the last is in sentence form, becomes confused by the nonsentence structure of the last. Likewise, things given in phrases or mere word groupings should be alike in their structure from one to the next.

In an outline, entries should be either all words, all phrases, all clauses, or all full sentences.

Dosage 41: Paralleling

Rewrite each of the sentences below to correct the nonparallel structures.

He added salts of calcium, sodium, and sal ammoniac.

Activity was detected in regions *b* and *c* not only when antisera to HSV was reacted with infected cells, but also when normal sera were substituted, and to a limited extent in reactions involving uninfected cells.

Some techniques that can be used to teach the student the art of medicine are to: (1) have him observe the way the preceptor handles problems; (2) observe decisions of when to hospitalize patients; (3) how to use community resources.

He was graduated at the top of his class, president of the student body, editor of the school's publication, *The Zinger*, and will be awarded the Fernal Foundation Award in September.

---

13. Paragraph Sense

Sentencing was so simple, so familiar, and so common sensical, that you will be cheered to find that paragraphing is no more complicated, no more foreign, and no more a departure from coordinated thinking than the way you already approach science, research, or clinical practice. You will even see ye olde scientific method within the paragraph.

PARAGRAPHING

It behooves us to know that the word "paragraph" derives from the Greek *paragraphos*, the name of a short line in a text that marks a break in sense. The paragraph is a means of showing the step-by-step development of the writer's message, each paragraph reporting a step. In a science article, especially, the reader relies on the logical step-by-step progression in "reading the message," whether by carefully reading each section, skimming paragraphs from section to section, or merely looking for "key terms" in dancing his sight down the columns. Paragraphs are minimessages in themselves, each telling a critical, integral part of the overall message, an effective grouping of statements about an important point to be made in the context of the whole. Each paragraph should have a complete message and internal movement; a group of paragraphs, as a block, likewise should have a complete, larger message and overall sequence.

Movement

How does one sentence in a paragraph relate to the next? What is the sequence of sentences? As in all forms of writing, such as news stories, editorials, and human interest stories, paragraphs in science writing can be developed in several ways, that is, the sequence in which the information is given can vary, depending on the purpose.

Once again, if you can see beyond the specific messages of the cells in the paragraph, you will perceive the overall structure of the para-

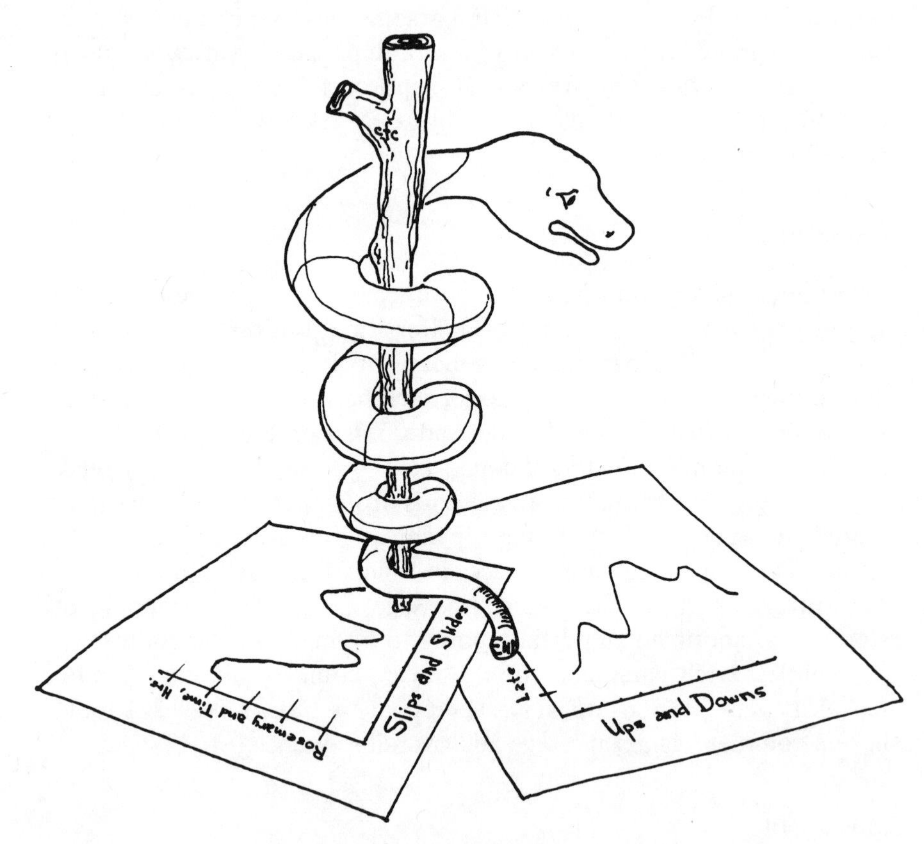

Fig. 13-1. I thought he said "pair o'graphs."

graph and how that structure moves from the beginning to the end of the paragraph. Then, how to add the second and third paragraphs will become clearly evident as the next logical step.

Four kinds of paragraph movement, or order, are basic to science writing:

(1) General to particular (deductive reasoning)
(2) Particular to general (inductive reasoning)
(3) Descriptive (parts of the whole, enumeration, or classification)
(4) Cause and effect

Deducting the Facts: The first category, general to particular, is nothing but the deductive mode of thinking you are familiar with in science and the order most typical of science writing. You take the main idea of the paragraph—perhaps it is a conclusion about your results—and you break it down into its specific components, showing the step-by-step debunking. That main idea in the paragraph is the topic of the paragraph, and the sentence that clearly states that idea is the *topic sentence.* If deduction is the movement of your paragraph, you should state the topic sentence as soon as you begin showing your deductive reasoning, because the logic of subsequent statements will flow from that basic idea. Usually the topic sentence is the first or second sentence in a deductive paragraph.

You might begin writing the paragraph by writing the topic sentence. Then ask: What subunits of thought does that idea comprise (include, cover, serve as an umbrella idea for)? Analyzing the idea deductively is similar to taking that first cadaver apart—systematically.

Example [*general to particular* (*deductive*)]:

The conducted responses recorded directly from the cord differed greatly from those recorded at the skin level. The simple monophasic potential recorded at the skin level is indeed associated with the polyphasic potential recorded directly from the cord though it may not be a direct manifestation of it. Comparison of the two waveforms illustrates that the surface response begins before the cord response beneath the skin electrodes. This implies that, in the volume conduction of this response, the surface electrodes are located at a point which acts as a source for a distant, but approaching, sink. As the response occurs beneath the skin electrodes, the rapid potential changes in the polyphasic potential are attenuated by the resistance and capacitance of the intervening tissue. Little indication of this activity can be seen at the skin level. A large slow wave due to the deeper sink is the only apparent feature.*

Dosage 42: General to particular

Write a paragraph that moves from a general concept to particular components from this information:

Smoking harms.
Sixty million American citizens smoke.
"Proving" a cancer association is difficult in humans.
Dependency may relate to smoking.
The "ritual" of smoking may be important to dependency.
Dependency pertains to alcoholism, drug use/abuse.

Inducting the Fact: The second category, particular to general, is the inductive mode of reasoning, the reverse of deduction, like putting the systems of the cadaver back together. One by one, in logical, progressive order, you state specific details or thought units and lead your reader step by step, reason by reason, to the greater idea. Your approach is, "Here are the specific reasons, what larger idea do they compose and conclude?" What do they add up to? In such a thought movement, the topic sentence might well appear at the end of the paragraph.

Example [*particular to general* (*inductive*)] :

The large segmental responses could be easily recorded either directly, by electrodes placed on the dorsum of the cord, or indi-

*Leo Happel et al., "Spinal cord potentials evoked by peripheral nerve stimulation," *Electroencephal. Clin. Neurophysiol.* **38:** 349-354 (1975). Quoted with permission of the authors and Elsevier Scientific Publishing Co., Inc.

rectly, by skin electrodes placed over the lumbar spinal column. These potentials are similar in waveform to those reported previously (Lloyd and McIntyre 1950). Common reference recordings reveal an initial sharp negative transient followed by a sequence of smaller, more prolonged positive and negative waves. Bernhard (1953) has shown that the later negative and positive potentials are related to activity in post-synaptic neurons, and are not conducted dorsal root potentials. Similarities between the responses recorded by both methods confirm the accuracy of skin-recorded responses as a measure of segmental spinal cord activity. The abolition of these responses after appropriate dorsal root section indicates that the potential recorded at both skin and cord levels is sensory in nature, and that antidromic discharge in motor fibers does not contribute significantly to the observed response. Thus, the sensory component of a peripheral nerve can be examined apart from the motor component by using this noninvasive method of recording the cord potential evoked by stimulation of the appropriate peripheral nerve.*

Dosage 43: Particular to general

Write a paragraph that moves from particular elements to a general concept. Use the same information given in Dosage 42.

Displaying the Facts: The third category of paragraph movement is less of a display of a reasoning process and more a display of details, a description of the components of an idea. Such a paragraph might describe the parts of a whole; it might enumerate or specify, for example, the results of an experiment; or it might describe the multiple steps in an experimental method. The topic sentence in such a paragraph would occur at the beginning and would be a general statement of the process or object to be described in the paragraph.

Example (descriptive paragraphs):

Spinal cord recordings were made in 20 cats. Six animals were

*Leo Happel et al., "Spinal cord potentials evoked by peripheral nerve stimulation," *Electroencephal. Clin. Neurophysiol.* **38:** 349–354 (1975). Quoted with permission of the authors and Elsevier Scientific Publishing Co., Inc.

anesthetized with sodium pentobarbital 30 mg/kg given intraperitoneally. Fourteen animals were given fluothane in oxygen during stereotaxic decerebration with a radio frequency lesion in the midbrain. In the latter group, anesthesia was discontinued approximately 30 min prior to recording. Rectal temperatures were continuously monitored and maintained between 97° and 100°F.

The left sciatic nerve was exposed in the thigh for stimulation with bipolar encircling electrodes. Supramaximal (3.0 V, 0.5 msec) stimuli were used to evoke the responses. In some cases stimuli were given at a repetitive rate of 3/sec, whereas in others the stimuli were randomized with successive pulses applied at 0.3-1/sec.*

How'd That Happen?: The fourth category is the gritty-gristly paragraph of the science article. In it, you start with a fact that is a result of your research, an effect for which you must explain the process by which the effect results, the how and the why. In the paragraph, the topic sentence is the statement of the result or effect. A hypothesis expresses such an inquiry into a cause/effect relationship, and, in wrestling through the process of unifying your conclusions, you have come to know this category well.

Example (*cause and effect*):

All the cells that met our criteria of "plasticity" did so whether whole trials or only the latter halves of trials were examined. This fact indicates that the differential response of these cells is manifested long before any overt behavioral response occurs, and provides reassurance that these findings do not arise from nonspecific sources such as movement, direction of gaze, and so forth. Similar reassurance is derived from the fact that 39 of the 56 cells examined showed no significant differences either early or late in the post-stimulus response. Perhaps the most convincing support for the statement that these findings are not attributable to nonspecific causes comes from the fact that 15 of the 16

*Leo Happel et al., "Spinal cord potentials evoked by peripheral nerve stimulation," *Electroencephal. Clin. Neurophysiol.* 38: 349-354 (1975). Quoted with permission of the authors and Elsevier Scientific Publishing Co., Inc.

plastic cells showed no significant differences during the early segment of the post-stimulus histogram. The latency span of the early segment of the histogram corresponds to the sensory-specific or exogenous portion of the evoked response, and should reflect changes in the afferent input that might derive from changes in the direction of gaze or level of arousal. All 16 plastic cells showed changes in the long-latency segment of the histogram that corresponds to the latency region of the endogenous components described in the introduction to this report.*

Telling the Tale: A fifth form, often used in popular writing but seldom used in science writing, develops the train of thought from chronological event to chronological event. The following paragraph is a *chronological narrative*, that is, it tells a story in time sequence.

Example (*narrative paragraph*):

The patient had been a farmer, owning about a dozen each of dairy cows and hogs. Brought to the hospital with severe tinnitus, he complained of having persistent diarrhea for more than a week. That complaint led to an examination of his seven family members and then to interviews of four neighboring families. All were found to have had diarrhea of various degrees of severity during a one-week period. Later contact with the county health department led to further studies and to the final accumulated total of persons involved in the epidemic: 709.

Many uninformed persons refer to the textual part of a book as the "narrative," even when the text does not tell a story. To narrate means to tell a story; a narrator is a storyteller and the narrative is the story told. The appropriate word for a text that does not tell a story, or even for a text that does tell a story, as distinguished from charts, tables, and cartoons, is "text."

A good narrative paragraph is sometimes appropriate in science writing, for example, in describing a patient's case history, or in telling the fascinating way you stumbled on the key to science you are

*Alexis Ramos et al., "Stable and plastic unit discharge patterns during behavioral generalization," *Science* **192:** 395 (April 23, 1976). © 1976 by the American Association for the Advancement of Science.

describing in your article. Such paragraph development usually leads to greater readability but also requires great care on the part of the writer not to sacrifice information merely to give the reader more drama.

Completeness

Do you know how much your reader already knows and what he needs to know to understand your idea? Deciding that will influence the amount of information you must put in each paragraph; however, the main idea in the paragraph and its integral development will be the most important determinants. Therefore, although in a science article most paragraphs contain several sentences, some paragraphs require more because the points at issue must be explained more fully than other points in the manuscript. Paragraphs usually have three to six sentences; however, the important guideline is not the length but whether or not a paragraph supplies all the necessary information.

The paragraph should have a sentence that states the topic you intend to amplify, support, or refute. That statement may be a question that you will proceed to answer. The sentence is called a *topic sentence* and usually is the first or second sentence of the paragraph. To facilitate your writing at this stage, begin your paragraph with the topic sentence. The following two paragraphs are the same except for their topic sentences:

(1) The original papilloma is not known for certain, but several hypotheses exist. Ulmann suggested that a virus causes the laryngeal disease, which then transplants down the respiratory tract. Others have speculated that the origin may be multicentric, that the disease may result from trauma, or that the implantation vector may be the air. Thus, both origin and means of spread are undetermined.

(2) What is the origin of papilloma, and how does it spread? Ulmann suggested that a virus causes the laryngeal disease, which then transplants down the respiratory tract. Others have speculated that the origin may be multicentric, that the disease may result from trauma, or that the implantation

vector may be the air. Thus, both origin and means of spread are undetermined.

Dosage 44: Topic sentences

Underline the topic sentences in each of these paragraphs; then determine whether each is deductive, inductive, descriptive, or cause and effect; finally, enter in your notebook a one-sentence summary of the gist of each paragraph.

(1) Not long ago I visited an art museum in Texas. Mounted on one wall was a beautiful Maya stela bearing a huge figure in full regalia flanked by glyph panels. Its looters had "thinned" the monument to a sheet of stone only an inch and a half thick, then sawed the sheet into smaller squares, damaging parts of the sculpture. Where had it stood? Only its looters knew.*

(2) In a second experiment, subgroups of Jewish and Protestant *Ss* were told either that their religious group typically takes less or more pain than other religious groups but in this case an explicit comparison was made between Jews and Christians. Both Jewish and "Christian" *Ss* increased their pain tolerance when told their groups were typically inferior in regard to this variable. The Christian *Ss* who were informed that their group was superior in pain tolerance futher increased their tolerance while Jewish *Ss*, similarly treated, showed no reliable change in their tolerance levels. The findings are conceptualized in terms of a theory of membership groups.**

If you find that the supportive statements in a paragraph are so numerous that the paragraph extends beyond what *two* paragraphs would normally take up in space, perhaps even beyond a column's length, such that the paragraph begins to become "visually burdensome," it is a good idea to analyze your statements for affinity. Per-

*George E. Stuart, "The Maya: Riddle of the Glyphs," *National Geographic* **148**: 768–791 (Dec. 1975).

**Wallace E. Lambert et al., "The effect of increased salience of a membership group on pain tolerance," from *Current Studies in Social Psychology*, edited by Ivan D. Steiner and Martin Fishbein. © 1965 by Rinehart and Winston, Inc. Reprinted by permission of Holt, Rinehart and Winston.

haps they really explore two slightly different, though closely related, aspects of an idea and can logically be separated into two paragraphs.

Dosage 45: Paragraph completeness

(1) An excerpt from an article in *American Scientist** is printed below with all the paragraphs run together. Analyze the text for completeness of ideas and use the paragraph sign (¶) to break the text into paragraphs. The original text has seven paragraphs.

Scientists and the Policy Process

What then can those of us who are scientists and engineers do to facilitate the several roles of science and technology in the more effective promotion of human welfare? Our effort like our roles must, I think, necessarily be multidimensional. It requires, first of all, a serious individual commitment to public life on the part of large numbers of scientists and engineers: many of us must seek active roles in policy-making both in the public and in the private sectors. *This includes seeking public office.* It is a serious indictment of our community that the number of scientists and engineers in the Congress literally can be counted on the fingers of one hand. It was Pablo Neruda (1974), the distinguished Chilean poet and Nobel Laureate, who said in his *Toward the Splendid City* that he learned to write poetry not by studying poetical forms but by experiencing the redeeming features of human nature in the selfless acts of strangers. We in the scientific community must similarly learn the humanizing insights that come through other people and the essential value of a wider sense of communal understanding. "No poet," says Neruda (p. 21), "has any considerable enemy other than his own incapacity to make himself understood by the most forgotten and exploited of his contemporaries, and this applies to all epochs and in all countries." I would hope we would recognize the virtue of substituting *scientist* for *poet*.

*William Bevan, "Science in the penultimate age," *Amer. Sci.* **65**: 538–546 (Sept.–Oct. 1977).

But individual commitment, while necessary, is not a sufficient prerequisite to the resolution of problems of the immensity of those we must now confront. Action must in one way or another be institutionalized at the local, at the national, and ultimately, at the international level. A relatively young organization, Common Cause, with a current membership of not more than several hundred thousand, is becoming increasingly effective in its representation of the public interest to our national government. There is ultimately no reason why the scientific and technological community, with its roughly 2 million members (Rose and Rose 1975, p. 20), cannot, by working through instruments like Common Cause, achieve a greater effectiveness in relation to matters of public interest than is now the case. Certainly, I think that it is imperative that the scientific societies take more and more effective public roles than they are now taking. Of course, one serious problem associated with any direct involvement of the scientific and engineering societies in the lobbying function is that they are prone to confuse guild with public-interest issues. At the same time, the international scientific community, because it indeed has already made great progress toward transcending national boundaries and constraints of perspective, has enormous potential as an effective force for good in tackling the problems that trouble us on a worldwide scale. . . . We in the schools, colleges, and universities have a special responsibility to the future effectiveness of science and technology. We must educate for a wider spectrum of societal roles. We must rise above the parochialism of particular disciplines and work settings. And above all, as Richard Atkinson, Director of the National Science Foundation, has recently pointed out (1975), we must find ways of breaking the bottlnecks within our own institutions to the career advancement of vigorous and talented new recruits to the scientific and technological professions. Whatever our ultimate strategies, we must now join others in insisting on the development of policies that provide for *integrated* planning. The need for such planning is a matter of conscious conviction with conservative and Marxist alike. . . . But to recognize the need for planning is not enough; we need strategies for planning that work. No nation, East or West, has so far utilized any that have worked well—or even adequately for that matter—for any sustained period of time. Certainly as we look to planning we must keep in mind the full scope of the scientific and technological enterprise. We academics are inclined to talk about the future of science most often in terms of basic research or university-based science. But these represent respectively only 8% and $12\frac{1}{2}\%$ of the nation's total R&D expenditure (NSF 1974), which now rapidly approaches $40 billion per annum from all sources. Indeed, in the 1975 fiscal year about 90% of our expenditures for science in this country went for military R&D or for research associated with the production of goods. . . . Finally, while I believe that we urgently need better public policies with regard to the societal roles of science and technology, I am quite ambivalent about whether or not the government should be the major agent in the execution of these policies, for I share Peter Drucker's intuition that governments are probably only effective at waging war and inflating currency.

(2) The following two paragraphs are reproduced *as published*. Can you break them into more paragraphs and maintain completeness in the paragraphs?

The Miners: A Special Case

To the miners' weapon of an overtime ban, the Government has replied by taking emergency powers and imposing a three-day week throughout industry. The time is now ripe for suggestions of Governmental retreat with dignity, and a formula for this has been devised by Dr. Anthony Freeman in a letter to *The Times*. He proposes that the offer to the miners could be increased, without violating the Government's wages policy, by large increases in compensation for the injuries and ill-health particularly associated with mining. These are indeed formidable. Risks of death from industrial injury are 10 times greater than in manufacturing industry. Miners' standardised mortality from all causes is now 15% higher than the mean for all occupations, compared with 6% higher in 1931. Their wives have a 29% excess, showing that social factors, outside the work itself but connected with the entire social milieu of coal mining, are an important determinant of illness and its outcome. Earnings fall with increasing age and experience, as the miners become less able to cope with the extremely arduous nature of work at the coalface. A collier on a fully mechanised face still has an average energy expenditure per shift of 1790 Kcal., compared with 1990 for hand-hewing, and in most pits there is still a great deal of shovelling, often from a kneeling position. Even on a "button job" a man may have several miles to walk underground, with steep gradients, over irregular ground, and often a low roof. In one South Wales mining village, 25% of men aged 35–54, and 61% of those aged 55–64, were found on clinical assessment to have important chronic disabilities affecting employment. The absence of suitable alternative work for the disabled in colliery districts is notorious, and most are also areas of high unemployment. Inception-rates for incapacity in miners are 182% and average days of incapacity 217% of the England and Wales mean, and general-practice consultation-rates are 124% of the mean. These figures represent real disability; studies of the proneness of miners to complain of rheumatic pain, compared with other manual workers and office workers, showed that miners had more advanced radiological joint damage at earlier ages, and that their consultation-rates, though high, were proportional to the objective evidence of joint disease. No association was found between psychometric test scores and rheumatic symptoms.

Turning to specific diseases, there are the vexed questions of disability from pneumoconiosis, and the contribution of coal dust to bronchitis. Simple pneumoconiosis seldom causes severe disability, but evaluation of lesser degrees of disability is difficult because of the crude nature of tests of respiratory function normally available, with ranges of normal that vary widely between individuals. Even more difficult is any attempt to estimate disability chiefly from radiological change alone, and it may be unfortunate that discussion of the subject has been so dominated by this sometimes deceptive variable. Necropsy studies correlating morbid anatomical change at a histological level with previous radiological category and tests of respiratory function have confirmed that function may be impaired in simple pneumoconiosis and that this impairment may correlate better with histological change than with radiological category. American work has suggested that evaluation of respiratory function in simple pneumoconiosis is difficult without assessment of gas exchange. In most disabilities the patient's reported symptoms are a more sensitive indication of departure from his own normal state than any objective tests, and the elimination of subjective data in compensation situations is an important handicap in accurate diagnosis; colliers rightly fear pneumoconiosis, and there is no evidence that they seek the diagnosis on any substantial scale. Complicated pneumoconiosis (progressive massive fibrosis) is beyond any doubt a crippling disease, and the B and C categories cause the same prolonged and very distressing terminal illness as do other obstructive respiratory diseases, with cor pulmonale and a fight for each breath that may last for years. It is now rare in young men, and in the elderly it grossly impairs life without very greatly shortening it, so that (as in bronchitis) mortality studies are of little value. Cochrane's claim that category A does not shorten or seriously impair life conflicts with a great deal of clinical experience, and has been criticised on methodological grounds.*

As every writer soon learns, completeness may be met in a single sentence. Often a writer with limited experience, having concluded a manuscript with a final thought, is reluctant to leave it at that—a singular, all-inclusive, apt, succinct, nutshell statement. Such nutshell paragraphs in science writing, uncommon as they are, nevertheless have their place.

Examples:

With the requisite sensors and effectors installed in the household, the public-utility information system will shut the windows when it rains.**

*"The miners: A special case?" *The Lancet* (*Editorial*) **1**: 81 (Jan. 19, 1974).
John McCarthy, "Information," *Sci. Amer.* **215: 71 (Sept. 1966).

> If we grant the existence of black holes, it is natural to inquire next as to what types can occur. In other words, what are all the solutions of Einstein's field equations which describe black holes?*

Unity

Unity that results in consequence is the logical quality of the paragraph, the internal relatedness of all its sentences to one purpose. Whatever may be contained in the paragraph that does not contribute to its singular purpose distracts from that point and may affect the reader's understanding of the paragraph's message.

Consider whether or not this paragraph has unity:

> In most cases of lapsus linguae, the sequence of symptoms is sinister. The victim gradually loses composure, develops increasing facial color, and begins to show minor movements of increasing frequency ("fidgetiness"). If mental retardation is a factor, neither Homans nor Cullen has reported any evidence for it. The eyes may show lateral or vertical motion, or both, and the buccinator reflex may be evident.

The comment about retardation does not add to the purpose of delineating the "sequence of symptoms" and does not belong in the paragraph.

A paragraph retains unity as long as the multiple points or ideas evolve from or revolve around a principal purpose.

> Two methods are used, the first equally as good as the second. Although the "glottis mobility encounter" (GM) technique requires no equipment other than a tongue depressor or clean finger, the "tongue-control frequency count" (TC) is measured with a Daedelus counter, and hence is more costly. Neither method has yet been evaluated in a study of more than 12 patients. GM, however, has the approval of the Transvaal Royal College of Oral Urchins.

*Robert Wald, "Particle creation near black holes," *Amer. Sci.* **65**: 586 (Sept.-Oct. 1977).

The writer could continue with discussions of each method separately, or, if the paragraph were sufficient to the need, he could discuss the methods or applications.

When writing an article, a scientist may well come to the dilemma of having to bifurcate his thesis to discuss first one aspect of it and then, presumably later, the second. He may be tempted in such circumstances to develop his paragraphs with the first line of discussion entwined with the second. Such a development, with Argument A mixed with Argument B, if it has possibilities at all, is usually brutal for the reader and futile for the writer. A form more likely to be understood would be one in which the writer explains, in an opening paragraph, that first the one line of argument and then the other will be presented, followed by a comparison of the two (if appropriate).

Example:

> If the major part of the conducted, whole-cord potential is mediated by non-primary afferent fibers, it reflects neurally coded transmission. The potential recorded from non-primary fibers is related to the probability of discharge determined by the code. Thus, successive responses to a repeated stimulus may differ slightly, according to probability. The coded response could have a variable latency to successive stimulus pulses or might demonstrate a complex pattern of responses to the individual stimulus pulses. Averaging these coded responses could produce the waveforms described in this study.
>
> An alternative explanation may also apply. The polyphasic nature of these potentials might represent cord conduction at different velocities through various types of fibers. This possibility is substantiated by microscopic observations that ascending tracts are composed of heterogeneous fiber sizes. Furthermore, the existence of many different ascending tracts supports the tenet that differential conduction velocities account for the polyphasic-conducted potential. The use of large-diameter electrodes resting on the surface of the cord would allow sampling of electrical activity of a large portion of the volume of underlying tissue, and not only potentials confined to the dorsal column system.*

*Leo Happel et al., "Spinal cord potentials evoked by peripheral nerve stimulation," *Electroencephal. Clin. Neurophysiol.* **38:** 349–354 (1975). Quoted with permission of the authors and Elsevier Scientific Publishing Co., Inc.

Dosage 46: Completeness and unity

Analyze the paragraph below for completeness and unity.

- Is there a topic sentence?
- What is the topic?
- Do all the sentences develop that topic?
- Are any of the sentences or clauses superfluous (redundant)?

This dictionary has a vocabulary of over 450,000 words. It would have been easy to make the vocabulary larger although the book, in the format of the preceding edition, could hardly hold any more pages or be any thicker. By itself, the number of entries is, however, not of first importance. The number of words available is always far in excess of and for a one-volume dictionary many times the number than can possibly be included. To make all the changes mentioned only to come out with the same number of pages and the same number of vocabulary entries as in the preceding edition would allow little or no opportunity for new words and new senses. The compactness and legibility of Times Roman, a typeface new to Merriam-Webster dictionaries, have made possible more words to a line and more lines to a column than in the preceding edition, and a larger size page makes a better proportional book.*

Paragraph Clarity: Sequencing the Movement

A paragraph may have all the information required for a complete idea, it may have no extraneous information and hence have unity, and it may have an overall movement of presentation and logic; but, to have clarity of expression, it must also have appropriate sequencing of the sentences, transitions between the sentences, and thematic emphasis.

Sentence Sequence: Once you decide on the most suitable movement for the topic of your paragraph, you have some choices open to

*By permission. Preface to *Webster's Third New International Dictionary* © 1976 by Philip B. Grove, Editor in Chief, G. & C. Merriam Co., Publishers of the Merriam-Webster Dictionaries.

you about how to pattern the logic within the overall movement. For example, if you use category three (Descriptive Paragraph) to explain something, you must decide on the order of that description. Is it chronological–first step first, on to the last? Is it from top to bottom of a diagrammatic flow? left to right of a slide? symptom to symptom of the patient? Should you progress from head to toe? inside to out? insignificant to dramatic? Maybe you want to develop your idea through an analogy or by comparison and contrast. Remember, the reader has never seen the terrain you are describing, so lead him carefully, making no assumptions about his knowledge of the important elements. Do not give him a confused maze; give him a straightforward path from beginning to end.

Dosage 47: Sequencing the movement

Read the following summary of a science article. What category of movement is used in paragraph two? paragraph three? Within each paragraph, what is the basis for the sequence used to prevent the ideas?

Summary

Lumbar, thoracic, and cervical spinal cord responses evoked by sciatic nerve stimulation were measured in 20 cats at the skin level and directly from the dorsal surface of the cord. Computer averaging techniques were used for both skin level recordings and cord surface recordings at all levels.

Recordings made directly from the cord surface at lumbar levels were large and were characterized by a large, initial, negative transient, followed by a more complicated and variable waveform. As recording electrodes were moved rostrally, the initial large spike decreased in amplitude and the duration and latency of the responses were polyphasic, of long duration and small amplitude. Deafferentation by posterior rhizotomy of all lumbar and sacral roots ipsilateral to sciatic nerve stimulation abolished the response at all levels. Thoracic cord section also abolished the response over the cervical cord.

Skin level recordings were of shorter latency than direct

recordings, especially at cervical levels. Response configurations were similar for both recording techniques at lumbar levels, but had different waveforms at both thoracic and cervical levels.*

Unsequential Consequence (Non Sequiturs): An important concern in sentence sequence in a paragraph is the occurrence of the non sequitur, a statement that may ostensibly proceed logically from the preceding statement but in truth does not. Ordinarily the explanation for the non sequitur is that the writer presumed the reader would understand the relationship of both statements either from experience or by presumption. Another cause might be the author's failure to understand the function and need of the transitional sentence showing the relationship twixt A and B. The effect without the transitional sentence is to give the reader a sense of going from A to X or even Z. Or, more serious, there may be a basic flaw in the logic of the reasoning.

In the following paragraph, all the sentences relate to the topic and follow one another logically.

Paper Mills

Public and campus concern with the term-paper industry has shifted rapidly from indignation to inattention, leaving the paper mills churning at higher speed. A recent study concludes that "students at every major university in the country have access to at least one commercial firm that offers a variety of writing and research services ranging from term papers on any subject for undergraduates to M.A. theses and Ph.D. dissertations." Prices are reported to range from a bargain basement $2 per page for an undergraduate term paper (very likely one of many photocopies) to $10,000 for a custom-tailored "original" doctoral dissertation. According to the study cited, demand far outstrips supply; sales are limited chiefly by the firms' inability to retain a sufficient number of "qualified" ghostwriters. Other evidence suggests that some term-paper entrepreneurs overcome this limitation by making use of unwitting labor: papers are stolen from departmental

*Leo Happel et al., "Spinal cord potentials evoked by peripheral nerve stimulation," *Electroencephal. Clin. Neurophysiol.* 38: 349–354 (1975). Quoted with permission of the authors and Elsevier Scientific Publishing Co., Inc.

offices and sold, with already-graded "A" papers commanding the highest rate.*

Consider now the lack of sequence or nonlogic that would be introduced were any of the following three apparently related sentences inserted in the paragraph before the last sentence.

(1) Only wealthy students would likely use such a service.
(2) Only students with low IQ's would be likely to use such a service.
(3) Students attend classes for one another and answer roll calls for one another.

Transitions: The clarity of a paragraph is enhanced and facilitated by a smooth flow from one sentence to the next. Certain words make sentences cohere, stick together. These are transitional words, antecedents with correct pronouns, repeated key words and parallel structures, and references to the subject in the paragraph. Properly used, they give a paragraph coherence.

You should use transitional words when you want to give an example (for instance, for example, to illustrate), show likeness (likewise, similarly), show contrast (but, however, on the contrary, on the other hand, instead), add a component to the idea (also, finally, moreover, furthermore, again, too, first, second), to indicate time or place (later, earlier, formerly, subsequently, simultaneously, above, below), and to indicate a conclusion or result (therefore, consequently, accordingly, as a result, in conclusion). Such words are appropriate within sentences between clauses and at the beginning of a sentence to indicate the shift in sense between sentences or between paragraphs.

Thememphasis: If the sequence of sentences within a paragraph and of paragraphs within a section is orderly, the reader will read your thought quickly and comprehend it easily, moving directly on to the next thought-unit, or sentence, then on through the paragraphs and the entire article. If you want him to stop abruptly and

*Amitai Etzioni "Paper mills," *Science* (*Editorial*) **192:** 22 (April 23, 1976).

do a "double take"—really notice something—devise the sequence accordingly. For example, you might lead him through a crescendo of thought ending on a strong, short, emphatic note—a simple sentence, well pruned. ("Slow-virus disease" misnames the virus and the disease.) On the other hand, you could state the important idea first, then progress in decrescendo fashion. Or, you might change the pattern of your usual sentence structure so that the sentence stops the reader and makes him reflect thoroughly (Are viruses either "fast" or "slow"?); or you might end a section, beginning the next with a subtitle (Viruses: (a) conventional, (b) unconventional). These formal elements have meaning and convey a message to the reader. Good writing, therefore, relies on more than word meaning for thought development and communication; it also relies on form.

PARAGRAPH BLOCKS

Opening Paragraphs

Some writers find the opening paragraph the hardest to compose because they know it is so important. If the reader does not begin right off to get the message, or if he finds the approach too general or too esoteric, he will be disinclined to proceed to the golden goodies stored for him in the later parts of the whole. A good beginning for most science articles is a direct one—whereby the theme is blatantly bared and after which the specific contents of the article can be revealed.

A truism in writing is that the writer's first paragraph, rife with pseudoprofundities and revelations of dramatic and glorious insights, should be discarded and used only as a "running start." After writing your Introduction, you should consider what value the opening paragraph has for the article. You may opt to acknowledge that it is merely a "running start," thank it for its unheralded service, and discard it forthwith.

In planning the opening paragraph, think about whether you are elucidating a cause for an effect, or vice versa; describing a process, a procedure, or an observation; or answering an important question. Such considerations should lead you to appreciate what the substance of the paragraph should be.

Internal Paragraphs

Once you master the order of paragraph movement, you will immediately see that a section of an article, for example, Methods and Materials, has the same possibilities of movement and that, in fact, the overall article has one of those movements.

Concluding Paragraphs

Many journals now consider the final paragraph or two the appropriate place to single out conclusions, speculate on causes, and comment on future intentions for extensions of the study. Great latitude is given to this section, although the content should have pertinence, reasonableness, and justification. Here is an example of an appropriate concluding paragraph for a science article.

> The theoretical investigations of the process of particle creation near black holes may provide physicists with a rare opportunity to make progress toward these goals. Already the investigations have suggested new limitations in the previously "known" laws of baryon and lepton conservation and a possible new law of physics, the generalized second law of thermodynamics. It is impossible to predict the future of physics, but it may well turn out that further understanding of the black hole will provide us with a key to further understanding of the laws of nature.*

*Robert Wald, "Particle creation near black holes," *Amer. Sci.* **65**: 585-589 (Sept.-Oct. 1977).

PART IV.
FORM AND COMPREHENSION: MIND FIELDS FATHOMED

14. Data Splashes

Today the works of Dickens are hard going; too much tundra covering the lode. But modern media lead us to lodes quickly, effortlessly, subliminally. If you would bore your reader, elicit a yawn from your listener, and bid still another's mind to wander, minimize the stimuli you foist on them. Contrariwise, if you would hold your reader's interest, inspire enthusiasm in your listener, and spellbind another, then appeal to their senses . . . give them stimuli. The graphic arts are part of the response to today's conditioned requirement for communication through stimulation.

Graphics, in the wonderful way these things happen, started out among the Greeks to mean "to draw" or "to write," and, after centuries of bits being drawn and written, today connotes "all the stuff except the text itself." Thus, graphics, a collective singular, singularly collects drawings, tables, charts, fancy lettering, and even the way the pages of a publication take shape. Obviously, the term impelled "photographic(s)" and that continuity would be better remembered today were we to consolidate to "phics," homonymous of "fix," the solution whereby the pix—as most tradesmen call them—are chemically stabilized.

Though words may have sufficed in the past to stimulate the reading audience, today's communicator needs to know of line art, tables, histograms, pie charts, pictograms, maps, grids, scales, slides, and (alas) camera-ready copy. Usually such phics are called "art" by those who work with them. In that sense, art, synonymous of graphics, is everything except the "copy," which is the typed text, the words you the writer have so carefully put to paper.

At some ebb or flow of the information with which you are working, you will decide that you must show the reader how your information sways the point and washes out all objections. Such information, which nearly everyone calls "data," may be a stack of returned questionnaires, a book of notes from the lab, or a computer printout. The data oblige you to take another step toward becoming the Compleat Communicator: to discover graphics.

Sometimes a picture may indeed be worth a thousand words. For example, in this text, particular note should be made of Figure 4-1 wherein the picture changes little but handily supports the text; the mental block(head) thinker (Chapter 3), who has no need for supportive text; and the drawings for Dosage 32 (Chapter 11) with which any text would probably be detrimental to the message.

However, were each picture worth a given thousand words, researchers and analysts would long ago have put aside the Thematic Apperception Test and Rorschach ink blots because no individual variation would be found. In science writing, and in science graphics, consistency of interpretation should be the objective: the same message should be derived by all who look at the picture. Lucidity and delight of insight should be the products of a decent exposure of your data.

You, Communicator, should learn to "see" your message in both words and phics, or representations, that we choose to call "data splashes."

TABLES

A tabular arrangement of data is customarily called a "table," though users of the term have not felt constrained by definitions of tabular arrangement, or even data. Even publishers are indifferent to the fine distinctions between a tabulation and a table and they pool all columnar splashes under the catchment of "table." One sees tables of contents, a table of logarithms, multiplication tables, water tables, and just plain lists.

The simplest table is a mere list of items. A list of contents, with or without corresponding page numbers, would, rightly or wrongly,

Table 14-1 Seldom Considered Conditions for Hypochondriasis: Ab Labels

Abasia
Abetalipoproteinemia
Abiatrophy
Ab-oukine
Abrachiocephalia
Abulomania

Table 14-2 Frequency of Use of Ab Labels in Munchausen Syndrome Dx

AB-LABEL CONDITION	FREQUENCY (PER 100,000 Dx)
Abasia	4.1
Abetalipoproteinemia	0.3
Abiatrophy	1.1
Ab-oukine	27.9
Abrachiocephalia	0.2
Abulomania	7.0

be called a *table* of contents by most people. A list of diseases and associated conditions may well be called a table. Table 14-1 is an example of a data splash that merely lists categorical information.

Usually tables are more than single lists; they show data for two or more foci of information, as does Table 14-2.

More patterns of organization are shown in Tables 14-3 and 14-4. Note that the disease entries could be easily extended in Table 14-3, but extension of Table 14-4 would push beyond the limits of a journal page.

Many other table models are possible, of course, limited only by your own persistence, the sensitivities of your publisher, and the patience and sanity of the typesetter. You should try out several forms for your table, considering the facility of comprehension and execution for the reader, your secretary, and the printer.

Before giving you some guidelines for preparing tables, it might be helpful to describe certain conventions of table form.

Table 14-3 Incidence of Ab-Label Conditions Among 47 Patients

AB-LABEL CONDITION	NO. OF PATIENTS		TOTAL PATIENTS
	MEN	WOMEN	
Abasia	1	9	10
Abetalipoproteinemia	7	3	10
Abiatrophy	14	22	36
Ab-oukine	6	32	38
Abrachiocephalia	14	0	14
Abulomania	14	33	47
⋮			

Table 14-4 Flip-Flopped Version of Table 14-3

	CONDITION					
PATIENTS	ABASIA	ABETALIPO-PROTEINEMIA	ABIATROPHY	AB-OUKINE	ABRACHIO-CEPHALIA	ABULOMANIA
Men	1	7	14	6	14	14
Women	9	3	22	32	0	33
Total	10	10	36	38	14	47

Table Conventions

The various items in a table have specific terms, depending on the way they are laid out or function in the table. Table 14-5 structurally depicts a table in anatomical terms, and Table 14-6 presents data in the pattern of the anatomy.

Table 14-5 Anatomy of a Table

(top rule)______________________________

	Bracer Head	
Category Head	Column Head	Column Head

(horizontal rule)______________________________

Span Line
Peg
Subpeg
Peg — Harvest
Subpeg

Span Line
Peg
Peg
Peg

Span Line
Peg
Peg
(bottom rule)______________________________
Footnote(s).

Table 14-6 Ab-Label Frequency in Diagnoses of Hypochondriasis, Malingering, and General Malaise

	FREQUENCY OF COMPLAINT (PER 100,000 Dx)			
	MALES		FEMALES	
CATEGORY (DEFINITION)	<21	>21	<21	>21
Hypochondriasis				
Abasia (inability to walk, defect in coordination)	1.9	2.0	0.1	0.1
Ab-oukine (yaws in Gabun)	0.0	0.0	0.1	0.2
Abrachiocephalis (anomalous absence of arms and head)	0.2	0.2	0.0	0.0
Malingering				
Abevacuation (evacuation that is either excessive or deficient)	0.1	0.1	0.0	0.0
Aboiement (utterance of barking sounds)	10.0	5.5	2.3	2.1
Malaise*				
Abiatrophy (premature loss of vitality)	0.4	0.3	0.4	0.4
Abulia (loss of will power)	1.2	1.3	4.9	0.1

*Omits diagnoses derived (paradoxically) on Labor Day.

Dosage 48: On the table

From Table 14-5, select the appropriate anatomical term for each of these terms from Table 14-6:

(*Example*) Aboiement ___Peg___

Malingering ________

1.9 ________

Females ________

Abevacuation	________
*Omits diagnoses derived	________
Category	________
Inability to walk; defect	________
Frequency of complaint	________

Gathering It Together: Table Manners

- Have all your data collected and analyzed before you begin to organize your table(s).
- Decide how much data to include. Usually you will want to show all the information that pertains directly to the point being made in the manuscript. If you have several points to be emphasized, you may want to consider preparing a table for each.
- Number the tables in the order they will be referred to in your text. Avoid assigning tables shared numbers, for example, Table 14-3A, Table 14-3B.
- State for yourself in one sentence the precise point you are trying to show with each table.
- Title each table. Clearly state what the table is about, omitting unnecessary words. Do not use full sentences as table titles. Titles for each table in the article should be consistent in style, capitalization, and use of abbreviations and numbers. If you use a subtitle, place it after a colon in the main title.
- Set the horizontal and vertical arrangement of your material by organizing your left-hand column first. It should contain the entries you want to relate to the other data in the other column(s). If the table does not seem to group the information clearly, you may have the data flip-flopped, whereby the column headings are in fact the information to be shown as the left-hand column and vice versa. (See Table 14-4, an example of Table 14-3 flip-flopped.)
- Arrange the entries in the left-hand column in a sequence, either by time, date, number, logic, or other element. Place a title over the column.
- Arrange the other column(s) to coincide with your left-hand column. Give each column a heading. If two or more columns are subunits of a larger concept, they may be "embraced" under a single heading. (See Table 14-5, bracer.) A line will suffice to show the relationship of columns under a common heading.

• If, in the organization of the table, you find that the arrangement must have major vertical subdivisions, use span-line headings (Table 14-5).

• Align the columns precisely. Usually the entries in the first column will be flush left. Numerals with decimals align on the decimals; those without decimals align flush right in the column. For numbers that can exceed 1, place a zero before the decimals. For numbers that cannot exceed 1 (*p* values, *r* values, caliber size) omit the zero before the decimal. Each numerical entry should be carried to the same number of decimal places.

• In the column heading, give units, percentages, and other information indicating the nature of the column entries. (See Table 14-2.) Such information is usually enclosed in parentheses.

• Where data are to be omitted from a "cell," the explanation for the omission should be made obvious to the reader either in the title, the column heading, or a special footnote. In the column itself, indicate an omission with double dots, dashes, or NA (not applicable).

• Special information pertinent to the table as a whole or to specific entries should be given in one or more footnotes. A single symbol may relate to all the successive entries in a column. Most publications follow this series of symbols for successive footnotes: *(asterisk*), †(dagger), ‡(double dagger), #(number sign), §(section symbol), ||(parallels), and then a repeat by duplication . . . **, ††, ‡‡, etc. The sequence of symbols relates to the subordinateness of the elements: title, column heads, span lines, pegs, column entries. The order moves from left to right and from top to bottom in the table. Some publications use lowercase letters, others numerals. If numerals are part of the table data, they should not be used to indicate footnotes.

Because the table should be a definitive source for limited information, anything pertinent to that information should be put in the table or explained in the footnote(s). An omission of patients or subjects should be accounted for; probability values should be given; types of tests used (t, χ^2) may be shown. If the table information does not agree with that given in the text, then you will want to clearly show in the table or text why that is so.

*Some persons not American-born might somewhat appropriately call the asterisk an "ostrich"; American-born persons often call it an "Askrish."

• Do not put too much data in one table. If your efforts result in a table having more than eight columns across, especially if they contain words and not just numbers or symbols, you may well find the printer objecting to its size. Check the journal or call the publisher; the experts will discuss the matter solicitously with you to avoid receiving a problem-engendering, monstrous table.

Special tables may need to be set broadside, that is, sideways on the page, or they may even continue for two or more pages.

• Use lines, or "rules," in tables judiciously. Each publisher has a house style for use of rules in tables, but most object to vertical rules and may object to anything more than top and bottom rules (Tables 14-1 and 14-2) except in more complicated tables (Tables 14-5 and 14-6).

• Special effects, such as color rules or shaded areas, might be considered to show emphasis or special relationships (Table 14-6). Again, if you see a need for such effects, ask the publisher for help.

• Type each table double-spaced. Use as many pages for each as necessary.

• Check to be sure each table is cited in your copy. Often the citation is in parentheses. Your text and table should complement each other both in theme and location in the text.

• For appropriate style for capitalization and other particulars consult the journal you will be submitting your article to. The text and possibly the journal's instructions to authors will provide samples.

• Use your tables as material for slides. Short tables can enhance oral presentations. But note that the absolute limit should be an area measuring ten double-spaced lines vertically and thirty-five characters horizontally.

In essence a table is a simple list or a composite of minitables. When a table begins to be too complicated, you can reduce it to two or more tables by extracting the simpler tables from the complex one.

Dosage 49: Separate tables, please.

Take this unwieldy composite of minitables and divide it into several tables.

DYSPLEXIA	SIGNS/SYMPTOMS	EXAMPLE	CORRECTION
Content Faults			
Pseudoerudition	Pretention, jargon, archaic terms	demise	death
Wordiness	Overlong sentences, dilution of meaning	The overall goal of this project is the development of a method of modification of human malignant cells.	Our goal is to develop a method for modifying human malignant cells.
Weak connective	Overuse of *and* or *but*	He was a Texan but had never eaten beef.	Although he was a Texan, he had never eaten beef.
Redundancy	Excess in words	blue in color; solid brass	blue; brass
Hedging	Too many qualifiers	Possibly his illness may due to . . .	His illness may . . .
Nonparallelism	Similar terms not having like structure	He learned how to open, to clamp, and when to close.	He learned how to open, how to clamp, and when to close.
Ambiguity	Two or more interpretations	He treated the patient with a cold shoulder.	He treated the patient who had a cold shoulder.
Weak verb	Forms of *to be*, *to have*	Most injuries are the result of carelessness.	Most injuries result from carelessness.
It . . . that clauses	"It . . . that . . ."	It is estimated by Grover that. . . .	Grover estimates that . . .
Verbing a noun	Forced verb form	Patients were autopsied	We did an autopsy on each body.
Passive overuse	Nonattribution of agent, lengthy sentences, "by" phrases	Such cases should be appreciated by the family physician.	The family physician should appreciate such cases.

Table 14-7 *(continued)*

DYSPLEXIA	SIGNS/SYMPTOMS	EXAMPLE	CORRECTION
Content Faults			
Problem of reference	Which, that, this obscure from antecedent	This leads to confusion.	This carelessness leads to confusion.
Jargon	Persistent use of "special" terms, often erroneously	By every parameter, the study was justified.	By every measure, the study was justified.
Stacked modifiers	Compounding and complexing of a term	Green monkey kidney cell transformation technique	A transformation technique involving Green monkey kidney cells.
Organizational Errors	PUBLICATION PROBLEM	GUIDELINE	RECOMMENDATION
Concept difficult in exposition	Too many words to explain concept	Use simple terms, simple sentences.	Augment text with drawings.
Overload introduction	"Chatty" intros use expensive space	Limit intro to 1–4 paragraphs.	Put pertinent history, comments in Discussion.
Duplication of data	Unnecessary table or text	Try tabulation.	Put all data into table(s) if possible.
Abrupt change in idea flow	Page make-up adds to confusion for reader	Include *all* steps in thesis development.	Use transitional statments to show the relation of parts.
Absence of abstract	Necessitates longer introduction, and a summary	Abstract should be 4 or 5 sentences.	State what you studied, what you found, and what you conclude.
Superfluous figures, tables	Take space, reader's time	Limit to 1–3.	Combine tables, figs.; use essentials only.

Grammer Blunders	EXAMPLE	CORRECTION
Noun/verb agreement	Each of them have one chance.	Each of them has one chance.
Sentence fragments	That 23 million adults have hypertension	The American Heart Association estimates that 23 million adults have hypertension.
Dangling participle	Sitting down, he noted the patient was more comfortable.	He noted that when the patient sat down he was more comfortable.
Telegraphing	Patient had spleen removed.	The patient had his spleen removed.

Misspellings	CORRECT FORM
principal/principle	principle/principal
pruritis	pruritus
caesarean	cesarean
consanguinous	consaguineous
homogenous/homogeneous	homogeneous/homogenous
dessicate	desiccate

In reviewing the journal you intend to submit your manuscript to, note in several issues the general length and number of tables. Most editors become sullen on seeing more than one or two tables in a manuscript. After you have prepared all of the materials for your article or book, consider whether each table is *essential* to your message; eliminate those that are not essential.

ILLUSTRATIONS

Despite the fact that you are not a draftsman, cartoonist, or commercial artist, you should not miss opportunities to splash your messages in pictures or to suggest to others (professional artists, publishers and their artists, and even printers) how the messages could be illustrated. First, it is easier to spot those opportunities if you appreciate that those who work with illustrations have many terms for referring to them—for example, cut, figure, drawing, graphic, art(work), chart, halftones, and engravings. You may refer to all of these by the generic term "illustrations."

The term "figure" is usually used to cite a specific illustration in the text. Some publishers prefer to cite tables and illustrations as "figures" and thereby use only one numbering system instead of a separate series for figures and another for tables. Refer in your text to the figure and give most of the pertinent information about the figure in the legend, the explanatory line beneath the figure. Avoid "see text" and similar expressions in the legend, and explain each figure fully and succinctly. Do not simply repeat information given in the text. Follow the style of your selected journal for indention and for use of abbreviations, capitals, and other points of style.

Plan it yourself. Think about your message in terms of appropriate illustrations. Are you using an apparatus in an unusual way? Can you best explain your procedure with a schematic? Would a bar graph show your results better than a table? Consider pie charts or pictograms, which use pictures to represent proportional results—the larger the picture the greater the result.

One planning procedure is to doodle and draw in pencil while preparing your manuscript. Ideas for illustrations often evolve simultaneously with the text. However, certain types of research reporting will require advance planning during the experimental or implementation phase of a project. For example, photographs may have to be taken well before the manuscript is underway.

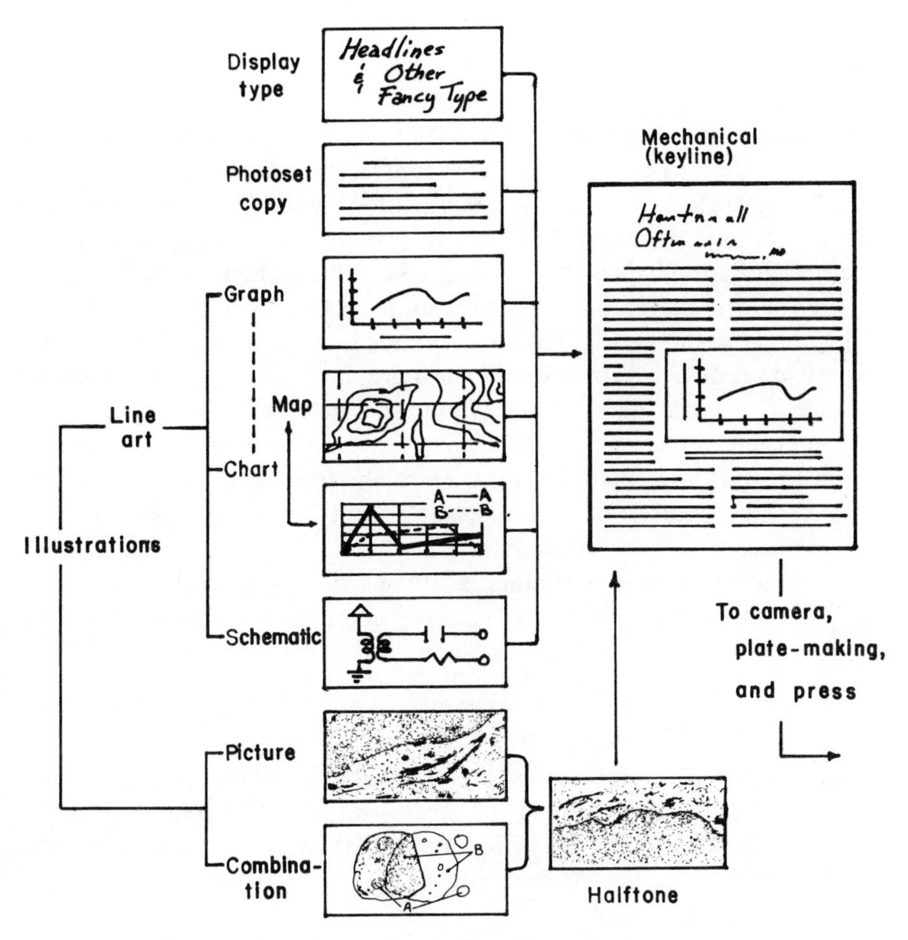

Fig. 14-1. "Pictures" are of many species.

Occasionally a publisher will provide graphics assistance for you, usually in preparing a book, but for a journal article you are pretty much on your own, and specifications from the journal are usually strict. Most medical schools provide medical illustrators and photographers, and it is a good idea to rely on these experienced artists to help you plan and execute your illustrations.

Photographs

The best prints from photographs are obtained from photos that were well shot to begin with. The contrast in photos, for example, should be sufficient to result in gradations when printed. Printers usually request high-contrast, glossy prints. The background of a picture should enhance the detail of the subject; hence, a solid light background is preferable for a dark subject and a dark one for a light subject. If your photograph has weak areas in contrast, exposure, or clarity of detail, a professional photographer can improve it somewhat with special darkroom techniques, but it is better to rely on proper camera settings in the beginning. (An out-of-focus picture cannot be put into focus by any trick of the technician.)

In taking pictures of patients, get permission from them to use the pictures with your printed material.

X-ray films present particular problems in reproduction. If you need to show such photographs, have a professional photographer reproduce the actual films as high-contrast, glossy prints.

Photomicrographs

Sometimes photomicrographs are erroneously called microphotographs. Again, these should be professionally prepared as glossy prints. Information on magnification and staining of specific points of interest is helpful. Arrows added to the prints should be solid black and outlined in white.

Color, or Hue Jest?

Because of the many problems involved, most publishers view the printing of color pictures with a jaundiced eye. Many journals do not use color at all. Nevertheless, occasionally color is appropriate for conveying the message of the illustrations in an article. Color can highlight whole or parts of tables, charts, and figures.

However, costs are usually a major concern with color production. Many publishers will permit use of color when it is essential to the message of an illustration, but you will usually be required to pay all or some of the costs. Generally the costs to you, whose intent it was to add some color to the lives of the readers, will be several hundred dollars, although the hundreds can sum to thousands with little difficulty, depending on the number of pages and figures involved, the policy of the publisher, and the circumstances that control the publisher's profit-and-loss statement. If advertisers flock to the journal of your choice and splash out expensive color ads, the production manager can reduce your costs by running the color ads on the same page-forms as your color figures.

Sometimes a second color is used to "fancy up" the appearance of journal contents, whence tables are enclosed in color lines, black-and-white pictures have a color tint, little color "artifacts" float around on pages, and even titles may be printed in color. The hand of the graphics expert is often surreptitious; the least troublesome attitude for you to assume is one of having no favorite color and of considering any color surprise a compliment to you and a complement to your article.

Slides

As soon as others learn that you have some data on a "hot" topic, they will want you to give a talk on it. Program chairfolks are for aye in search of "experts" to place on programs.

Slides can easily be made from parts of your tables. Here are the commandments on slide-making from copy or tables.

I. Always double space.
II. Type with sharp, clear letters; use capitals and lowercase letters.
III. Do not allow more than ten double-spaced lines, including the caption, and do not allow more than thirty-five characters and spaces in any line.
IV. Have the material professionally photographed and prepared as slides.

Other material and special effects may be added to your slide series to put some zip into your talk. Reversing the type, using color to show the progressive stages in the evolution of your data—these and other effects should be the responsibility of a professional photographer. Those who hear your talk will appreciate your efforts to illustrate the data for them; however, too many slides, especially when shown in the dark, can be as fatiguing and boring as an unillustrated talk.

15. Anatomy of the Article

WHAT/WHAT NOT

The disparity between the content and purpose of science articles has led in recent times to a difference in the forms that the articles take; but, clearly, one basic form of article has evolved for science journals. That form, established in American science journals more than sixty years ago, became so rigid that few questioned its logic, purpose, function, and appropriateness in light of continuing economic difficulties. Because of the questioning nature of science and scientists, such acceptance is surprising. What is not surprising is that today journal readers and editors, with more materials than time or space, have begun to put the form under the glass for analysis.

IMRAD is the acronym for that traditional form: *I*ntroduction, *M*ethods, *R*esults, *a*nd *D*iscussion. It has served as a criterion for comparing one article with another, especially for length and overall heft. And it has provided a standard and accepted guide for transmitting information to those who, being carried by the trend toward specialization, needed a familiar and functional form. Readers who encounter the form know what to expect. Anything not in that form is "alien" in the science milieu—editorials, commentaries, letters, humorous accounts of poignant experiences in the laboratory, budget reports, proceedings of committee meetings, and news notes on visitations, awards, appointments—however informative, pertinent, relevant, seminal, or valuable it may be.

Today, despite many efforts to alter it, the accepted form still reigns; long live IMRAD (apparently).

In the dissection of the article, we shall here consider the article's body to be that of the standard IMRAD model, later giving comments on the anomalous forms that exist and seem to be a-borning.

TITLE/SUBTITLE

On a scale of difficulty, writing the title comes close to the top and should probably be left to the last. Authors too often are cavalier about titles, apparently not realizing that the words therein direct the reader into the content, key the content for indexes, and show the relatedness of the article to other articles.

The first rule of title-writing is this: Tell as much about the article as briefly as possible. For those who cannot cut through to the heart of the matter in a single stroke of the pen, we suggest the following operation.

- First, write out all the terms pertinent to your theme. From that list, try various forms of titles, without any concern for length.

  Example: Sensitized Lymphocytes
  Antibody Titer
  Transformation
  Candida Antigen
  Migratory Inhibitory Factor
  Skin Testing
  In Vivo Production

- Then begin to arrange the terms in whatever order seems most logical. Ultimately you will come out with a title that includes all the essential referents to your article's message, except that, almost always, the title will be incredibly long.

  The Hypersentivity Skin Test by Means of Candida Antigen with Transformation of Lymphocytes, Production of the Migratory Inhibitory Factor, and Antibody Titer

- From that boggle, begin removing any words that are not really needed: "Means of."
- Next consider which terms can be reduced by rearrangement of the words: "Transformation of Lymphocytes" converts to "Lymphocyte Transformation." From that point, the shifting of the terms into the optimal placement becomes a matter of trial and error.

  Candida Antigen for Hypersensitivity Skin Testing

In that trial and error, you should seriously consider leaving in the main title only those terms that are "direct" to the article message. The other information can be put into a subtitle or perhaps omitted altogether. Indexers, and other users of such information consider the title and subtitle as one element.

However, in your efforts to be as brief as possible, do not be so brief as to be cryptic or misleading. An article titled "Sex and Socioeconomic Status" has as its first subtitle, "Sex as a Risk Factor in Epilepsy," which would imply that if you were born either a male or a female, you, de facto, have a risk factor for epilepsy. In fact, however, two columns into the article, the stated conclusion is that sex, either male or female, is not a risk factor for epilepsy. In the following newspaper* headline, who did the slaying?

Teacher Slaying Suspect Caught After P. G. Chase

- In writing the subtitle, do not connect it with the main title except as an appositive, with a colon ("Eyeblink Inhibition by Monaural and Binaural Stimulation: One Ear Is Better Than Two"). In other words, delete such connectives as "with," "without," "among"; the reader can appreciate the relationship between the two without the (now somewhat awkward) connectives. Consider this example:

  (Title) Comparison of Guggle and Zatch
  (Subtitle) With Report of 45 Cases

  The "with" connective actually misleads somewhat in that the reader sees the comparison as having relation with the report of cases, when in fact the report is *of* the comparison, not *with* it. A better form might be "Comparison of Guggle and Zatch: Report of 45 Cases."
- Finally, do not title your article with words that do not pertain to the article, that is, words that you consider a prelude to your cleverness or good humor sense. Let the editors themselves pontificate "From Olympus," "On Discovering Discovering," or "Zen and You and the Lamppost."

Most journals have strict principles for the acceptance of titles, and criticism of titles is an important part of the review process. You should know, however, that the title and subtitle introduce the reviewer to your work; a poorly conceived title sets the tone for heightened criticism by the reviewer.

**Washington Post*, October 3, 1976.

Dosage 50: Your title, please.

Apply the above considerations to the following titles and put them in better shape.

(1) Pathogenesis of an Unexpected Sudden Death
(2) Enhanced Diagnostic Power of Exercise Testing for Myocardial Ischemia by Addition of Postexercise Left Ventricular Ejection Time
(3) Hemodynamics at Rest and During Supine and Sitting Bicycle Exercise in Patients with Coronary Artery Disease
(4) Cardiac Death in the First 6 Months After Myocardial Infarction: Potential for Mortality Reduction in the Early Posthospital Period
(5) Identification of Sudden Death Risk Factors in Acute and Chronic Coronary Artery Disease

THE INTRODUCTION SECTION

Once upon a time, when science writers were (or expected to be) obliged to provide a comprehensive background for the reader on the topic in discussion, the Introduction often resembled a review article. Depending on the breadth of the topic, the Introduction could span the 6,000 years of written communications. The Introduction usually raised points that rearose in the Discussion section, and perhaps even in the Methods and Results sections.

Now the publishing adage is "Half as long is twice as good," and the extravagance of long Introductions has been curtailed. The curtailment spares the reader repetition with the article and repetition in general. Is it possible to read any article on atherosclerosis these days without being told straightaway that heart disease is the number one killer and that Viet Nam soldiers in their late teens were found to have atherosclerosis?

The Introduction can be a difficult section to write unless you understand its purpose and its appropriate qualities.

- *Brevity:* The reader must be given only a "handshake" with the topic and be prepared to receive the message of the article through a minimum of words.

- *Information:* The Introduction prepares the reader for the message of the article. He must therefore be given enough information to render him capable of understanding the message in a particular context.
- *Attractiveness and Interest:* The Introduction should interest the reader enough to persuade him to read the rest of the article. Interest can be sparked by a sense of discovery or learning for the reader.
- *Nonsummary:* The Introduction is not a summary of the article. Do not give facts or conclusions that are presented in other sections of the article.
- *Theme:* The Introduction should state the major theme of the article and may even do so in the form of a question that states the hypothesis.

In most journals, good Introductions extend no more than four paragraphs. Many journals prefer to have the Introduction limited to a single, succinct, albeit stimulating paragraph. Clues to writing an effective Introduction come from many sources: articles published on the subject at hand, conversations with friends, correlates implied by material within the article itself, and a thoughtful review of the current state of the art.

You need not write the Introduction first—perhaps last is best. However, inasmuch as you began your experiment or study with a question and a hypothesis, and inasmuch as the Introduction should state that question and hypothesis, the Introduction might be the most logical starting point for writing your article. Write it initially without concern for length; you can trim it later. Try several different openings for it: who said what most recently? or, what has never been said? If all else fails, think of how you would begin telling someone at lunch about the message embodied in your article. Better yet, *tell* someone at lunch what your article is about—and then make a point to remember what you said.

Examples of Introductions:

(1) The diagnosis of tuberous sclerosis is not difficult when the classic triad of seizures, adenoma sebaceum, and mental retardation is present. In the infant or young child, skin lesions may

not be present, and further diagnostic studies are necessary to establish the diagnosis. Computer-assisted cranial tomography (CT) is an important new noninvasive roentgenographic technique that enhances diagnostic capabilities and delineates multiple factors useful in the management of patients with tuberous sclerosis.*

(2) Idiopathic thrombocytopenia purpura (ITP) occurs rarely during the course of chronic lymphocytic leukemia (CLL), non-Hodgkin lymphoma, and Hodgkin disease. The report of Ebbe et al in 1962 of five cases of CLL complicated by ITP verified the existence of a clinical spectrum of mild to severe hemorrhagic complications in this context. Treatment with prednisone or adrenocorticotropic hormone or both in that study resulted in a rise of platelet count and initial control of bleeding in three of five cases. Splenectomy was carried out in one case with no improvement and early death, and in the last case, prednisone was poorly tolerated and early death ensued. We are reporting our experience with three patients having severe symptomatic ITP that complicated the course of their CLL. In these patients, the ITP failed to improve on prednisone treatment but was successfully treated by splenectomy alone or in combination with immunosuppressive-antineoplastic treatment for more than 5, $2\frac{1}{2}$, and 1 years, respectively.**

(3) Minimum convex polygons have been used in various fields of biology to delineate groups based on variation in two physical characters. A minimum convex polygon is a polygon of minimal area with its vertices on data points and with no internal angles greater than 180°. They have been used very effectively to delineate lower and higher groups of vertebrates based on a log-log plot of body mass and brain mass. On a much finer biological scale they have been used for the morphological characterization of castes within a single species, e.g. polymorphism between the castes of bees, based on a plot of wing length and abdomen width.

*Gilbert I. Martin et al., "Computer-assisted cranial tomography in early diagnosis of tuberous sclerosis," *JAMA* **235**:2323 (May 24, 1976).

Robert W. Carey et al., "Idiopathic thrombocytopenic purpura complicating chronic lymphocytic leukemia," *Arch. Intern. Med.* **136:62 (Jan. 1976).

Two of the most basic properties of a virus that can be expressed as single numbers are its size and the size of its genome. Size of the genome "sums up" many other properties and activities of the virus. The size of the virus itself will be related to the structural complexity of the particle. Large viruses tend to have many components and complex structures whereas small viruses have few components and a simple structure. In this paper I explore the usefulness of polygons for delineating possible viral taxa, using up-to-date data on genome size and particle size and the families, genera and groups of viruses recently established by I.C.T.V.*

Dosage 51: Introducing . . .

Here are two sets of information. Consider the information and decide which comments and facts are appropriate for an Introduction and which ones should be deleted.

*Set 1***:

- –Effects of dietary bulk on serum lipids are not clearly known.
- –Vegetarians tend to have lower serum cholesterol levels than nonvegetarians.
- –Vegetarian diets are high in plant fiber.
- –Pectin has been found to reduce serum cholesterol levels in animals on high-cholesterol diets, and, in one study, decreased levels in men.
- –Serum cholesterol is distributed among different classes of serum lipoproteins.
- –Most studies of the cholesterol-lowering effect of pectin were of the induction of hypercholesterolemia.
- –During the reversal period, endogenous metabolism is the source of serum cholesterol.

*R. E. F. Matthews, "Minimum convex polygons for the delineation of possible viral taxa," *J. Gen. Virol.* **36**:41–50 (1977).

Paraphrased and reprinted from *Life Sciences* **16:1533–1544, Leslie M. Berenson et al., "The effect of dietary pectin on serum lipoprotein cholesterol in rabbits," copyright 1975, Pergamon Press, Ltd.

–Fourteen rabbits were fed high-cholesterol diets.
–Our report describes the effect of pectin on serum lipoprotein cholesterol during and after induction of hypercholesterolemia.
–Rabbits fed a hypercholesterolemic diet and pectin had significantly lower levels of total serum cholesterol than the controls.

*Set 2:**

PLATO, a computer system that provides self-paced instruction to students, displays on a screen 22 X 22 cm, material that may comprise text, drawings, and even color photographs. The terminal now being used with the system involves a plasma display panel having a grid of 512 by 512 electrodes at the intersections of which a neon discharge can be ignited or extinguished. Students use a special key-set to interact with the material, and they receive instantaneous aid for any difficulty and reinforcement for correct responses.

If the material is well designed it tends to be highly interactive and to require from the student frequent answers to questions, predictions of outcomes on experiments, and data interpretation. The users of PLATO range from those learning to read to graduate students in the medical sciences. Now used in universities, colleges, public schools, and other institutions, the system has 950 terminals. More than 3,500 hours of instructional material in 100 subject areas are available. Here we describe how PLATO is used in university science education and research.

THE METHODS SECTION

Most authors are comfortable in writing the Methods section of the article: they know full well what the experimental design was, what apparatus was used, what kinds and how many subjects were studied,

*A corruption of the Introduction from Stanley G. Smith and Bruce Arne Sherwood, "Educational uses of the PLATO computer systems," *Science* **192**: 344-352 (April 23, 1976). © 1976 by The American Association for the Advancement of Science.

and how the data were counted or obtained. The writing does not require creativity. It does require precision, order, and directness.

Your objective should be to provide a description of the materials and techniques used, in enough detail that a reader could do such a study or know how your method might differ from other methods. Determine whether your method is essentially a chronlogical progression of the steps, a description of equipment, or a logical arrangement of simultaneously occurring considerations. Accordingly, present your Method and Materials section in a chronological, expository, or logical order.

Here is a checklist.

- Summarize and discuss each separately:
  Experimental design
  Subjects, materials, equipment
  Method of data collection
- Explain any modifications of standard design, materials, or equipment.
- Simplify the section by citing reports in which the design, materials, or equipment used are fully explained.
- Account for everything pertinent to the design and the message of your article.

You might try writing the Methods section early in the experiment, even as you are setting it up, and it will be a *fait accompli* months later when the results are in and you report the study. Or, for a ready-made description of the methods, refer to the written protocol you developed for yourself and your lab assistants.

Dosage 52: Methods madness

Unscramble this scrambled Methods section, indicating at the left the order you think the information should follow*:

_Subjects were Japanese children.
_All were healthy.
_Each measurement was to the nearest 0.1 mm.

*Derived from Shozo Takai, "Principal component analysis of the elongation of metacarpal and phalangeal bones," *Amer. J. Phys. Anthropol.* 47:301–304 (Sept 1977).

_The Jacobi method was used for the computations.
_There were 33 boys and 33 girls.
_Eigenvalues and eigenvectors of the matrices were computed.
_Standardized radiographs of the right hand were taken.
_The calculations were from logarithms of the measured values.
_Covariance matrices for boys and girls were calculated.
_All children had ossified epiphyses.
_All had incomplete epiphyseal fusion in all bones on the radiographs.
_Venier calipers were used to measure the total length of each tubular bone.
_Subjects with apparent brachymesophalangia, brachytelephalangia, or other skeletal defects were omitted.

In the rare event that your message is itself the report of a new technique or method, the gist of the message is in the Methods section. Most journals are interested only in important advances in techniques, not in incidental twists or small refinements on old designs, which might be publishable in forms other than an article: Letters to the Editor, Brief Notes, New Methods, or New Products.

When the message is about a truly substantial innovation in methods, design, or procedure, then editors will appreciate it as such and accept it as a bona fide article. The form of such an article may well follow this outline:

Introduction
The Apparatus (Design, Technique, System)
  Unmodified
  Modified
Advantages/Disadvantages

THE RESULTS SECTION

A logical sequel to the Methods section, the Results section carries the core of the message of most science articles; it deserves more than the slap-dash plop-put of printout totals.

Once analyzed, data will usually lend themselves nicely to tabulation or to illustration. Rarely will *all* the data from the printout be pertinent to your message, so choose only those statistics or results that truly have value to the point of your article. Give as much information as is necessary for the reader to appreciate how you got your totals, statistics, and correlations. Columns of information obviously should be tabulated; comparative data may be shown in illustrations or tables.

Think of the tables and illustrations as substitutes for, or complements to, the text, not as reinforcements of data already stated in the text. In fact, the text should give little, if any, of the information found in the tables and figures, except to point out information highly specific to the conclusions.

The section should not include any comparisons, evaluations, or discussions included in the Discussion section that follows. Comparisons of data can be reported in statistical significances in the Results, but the evaluation of those differences should be given in the Discussion. If you find that you have "editorialized" the findings (stating that A is better than B) in the Results, consider what led you to such statements and develop your reasoning in the Discussion section.

As with other sections, the first draft of the Results should be written out as completely as possible so that nothing will be overlooked. In later drafts, prune the section, retaining only what is essential.

Perhaps the best advice for preparing the Results section is that it be reviewed critically by as many peers or su-peeriors as possible. An author tends to be so close to his work and the writing that emanates therefrom that often he cannot see subtle and even blatant divergences between facts and hypotheses. The more eyes that see the section, the greater your assurance that all the columns add up, the curve goes up and not down, and the risk factors do or do not imply diseases.

Dosage 53: Sults and results

Do one of the following:

(1) Write a short Results section on the consequences of having your car serviced.

(2) Introduce this statement into a peer setting (e.g., at lunch), observe the results, and record those results in your notebook.
The Statement: I am trying to find out who it was that got Senator Proxmire's Golden Fleece Award and then went on to get the Nobel Prize.

(3) Write a Results section from your own work.

THE DISCUSSION SECTION

Note that some journals call the Discussion section "Comment," ostensibly to avoid a sense of "conversationalism." Whichever, it is in this section that the writer truly becomes author–he compares or contrasts his findings with those of others, holds forth his arguments, and winningly shows the importance of his message.

Ask yourself, "What answer did I get and why is it important?" Some key approaches to prime your thought in writing the Discussion would be the following.

- *Compare:* Explain or account for the similarities between your findings and others' findings.
- *Contrast:* Show how your findings differ from what might have been expected in light of information from other sources.
- *Explain:* Tell why your findings were or were not as you inferred from your design. What confounds in the data were there?
- *Conclude:* End your section with the point you want the reader to retain most.

Here is a recently published Discussion section.

The tubules that constitute the axonemes of cilia, flagella, and sperm tails are morphologically similar to cytoplasmic microtubules. However, the arrangement ($9 \times 2 + 2$) of the tubules suggests that structural and chemical differences may exist between these tubules and cytoplasmic microtubules. Biochemical isolation of cytoplasmic microtubules of neurons reveals two species of tubulin. Similar biochemical studies of axonemal tubules of flagella also exhibit two species of tubulin, one of which shares physical and chemical properties with one of the tubulins from neurotubules. Tubulin from chick brain has also been shown to assemble onto flagella and sperm tail microtubules. We have now demonstrated that tubulin extracted from porcine or rat brain cross-reacts antigenically with components of axonemes of cilia in the tracheal epithelium of the rat.

Although this evidence does not establish the identity of the proteins of microtubules from different cells or even prove complete chemical identity of proteins with shared antigenic sites, it does reveal a measure of similarity between microtubules from brain cells and those of cilia of cells of the respiratory tract. This degree of conservation of structure of the tubulin molecule from cell to cell across species lines has been previously reported. The difficulty in isolating purified cilia from this epithelium, which is composed of several different cell types in varying proportions, may preclude more direct tests of their composition. To an even greater extent, the limitations of isolation techniques would impede analysis of basal bodies from the ciliated cells. Electron microscopy with enzyme-labeled immunoglobulins, which permits an approach to the composition of these cell substructures *in situ*, enabled us to demonstrate that basal bodies also bind the antiserum with specificity for tubulin.*

THE CONCLUSIONS SECTION

Often in the past, scientific conclusions were included in the Summary (or Summary and Conclusions), wherein the reader found new gristle to chew on. Then, some journals abandoned the Summary and adopted the Abstract. They then searched their corporate souls for a statement of overall projections from the data and called it the Conclusions section.

In pre-Abstract days, if the reader read the Summary before reading the article, he was surprised to find that the conclusions were not alluded to anywhere else in the whole of the work. Except for the Abstract, this should still be the case. Therefore, if you have given the message in the earlier sections of your article, and a Conclusions section would be superfluous and belaboring, be satisfied and end your article with the Discussion section.

*R. E. Gordon et al., "Electron microscope demonstration of tubulin in cilia and basal bodies of rat tracheal epithelium by the use of an antitubulin antibody, *J. Cell Biol.* **75**:586-592 (1977).

If, however, the whole message needs refinement and extension in the form of highlighting projections or implications, suggestions for new research, or simply a fresh synthesizing statement on what was found, then this section should be used as the repository for such information.

Dosage 54: Cuss, discuss, discussion

Do one or more of the following:

(1) Write a short Discussion section on your performance with the Dosages you have done to this point. (*Compare* the advantages or disadvantages of the use of humor throughout the book with the form found in ordinary writing books.)

(2) Write a Discussion section on writing Discussion sections. (*Contrast* the needs of the science writer today with those of yesteryear.)

(3) Write a Discussion to compare and contrast the words *compare* and *contrast.*

(4) Write a Discussion section for your own article.

THE ABSTRACT

Perhaps the notable change in the form of the science article during the past twenty years has been the evolution of the Abstract appearing before the main body of the article. Formerly, articles were constructed with the view that the reader would proceed from beginning to end and find a quick review (salted with some philosophy) at the end, in a Summary. In time, editors and publishers came to appreciate that readers search for the Summary first (after the title), thereafter deciding whether the article itself should be read. Today, the Abstract has supplanted the Summary in many journals.

Great rationalizations have risen about the "initial summary," as to what its purpose should be, what form it should take, who should write it, and how it should relate to the text itself. The best advice that can be given about the Abstract is this: Write what you *as a reader* would want to have available to you.

Ideally, such a paragraph follows the main line of the article and is not repetitious of the article's Introduction. In a standard development, for example, the Abstract would include four sentences, one for each section of the article: Introduction (hypothesis), Methods (how the study was done), Results (what was found), and Discussion (conclusions). In an editorial in *JAMA**, the following steps are recommended:

(1) Explain what the problem is and why you are interested in it.
(2) State how you studied the problem, and the methods you used.
(3) Report your principal observation(s).
(4) State your conclusion(s).

By following such an outline, you will not be led into writing "pap" Abstracts, those that give no information whatever—for example, "A study was undertaken, the data accumulated, and some interesting observations were made. Our conclusions are given."

If your Abstract has more than six sentences, it is probably too long. If it has fewer than three sentences, it probably does not say enough (or it may reveal to you that you have had nothing substantive to say). Be sure that key words given in the text are introduced in the Abstract, that your study problem is stated, your results and conclusions are clearly stated, and your final comment is terse but apt, and that the whole Abstract reads smoothly and coherently.

You will use the Abstract for many purposes: distribute it to friends working on similar research, submit it to program chairpersons for publication in programs, or give it to the grant-getting agent at your institution as a key to your interests. Many indexing services will pick up the Abstract after publication and use it verbatim in their products.

Examples of Abstracts:

(1) Diazoxide given for hypertensive crises caused severe complications in two patients. Hypotension, anginal syndrome, cerebral ischemia, and right hemiplegia developed in one patient, and myocardial infarction in the other.**

*M. Therese Southgate, "On writing the synopsis-abstract," *JAMA* **222**:1307 (Dec. 4, 1972). Copyright 1972, American Medical Association.
G. Krishna Kumas et al., "Side effects of diazoxide," *JAMA* **235:275–276 (1976). Copyright 1976, American Medical Association.

(2) We used electron microscopy to study *Entamoeba histolytica* trophozoites in rectal and liver lesions. The amebal surface had bleblike structures. Also noted were subplasmalemmal vacuoles and plasmalemmal extensions generally similar to the "surface lysosome" and "trigger" previously described. We found on serial sections that the plasmalemmal extensions were dendritic complexes enfolding the membrane of the subplasmalemmal vacuole and extending to contact host tissue. We postulate that membrane-bound cytotoxic hydrolases enter the plasmalemmal extension via the translocation of the subplasmalemmal vacuole membrane and thus are available to act at contact sites. The small blebs may provide an alternate method for membrane-bound hydrolase activity.*

(3) Serum lipid profiles of 3,446 (19% of population) children, ages 5–14 years, were determined in a biracial community (Bogalusa, Louisiana) as part of a program investigating the early natural history of atherosclerosis. Black children had significantly higher mean levels of serum cholesterol than white children (170 mg/dl vs 162 mg/dl, $P < 0.0001$). On the other hand, significantly lower levels of triglycerides were found in blacks than in whites (61 mg/dl vs 73 mg/dl, $P < 0.0001$). Girls had higher levels of triglycerides than boys in both races (blacks, 64 mg/dl vs 59 mg/dl $P < 0.0001$; whites, 77 mg/dl vs 69 gm/dl, $P < 0.0001$). The racial differences in serum cholesterol and triglyceride levels were even more apparent at the 95th percentile. The serum cholesterol level remained relatively constant in all children until ages 11 and 12 years, after which a slight reduction occurred. This reduction was more pronounced in boys than in girls. In contrast, a significant increase in the level of triglycerides with age was observed in all children except black girls, the increasing slope being most pronounced in white girls.**

*Jane E. Deas and Joseph H. Miller, "Plasmalemmal modifications of *entamoeba histolytica* in vivo," *J. Parasitology* **63**:25–31 (Feb. 1977). *Courtesy of the Journal of Parasitology.*

Ralph R. Frerichs et al., "Serum cholesterol and triglyceride levels in 3,446 children from a biracial community," *Circulation* **54:302 (Aug. 1976). By permission of the American Heart Association, Inc.

REFERENCES/BIBLIOGRAPHY

The reference list should include only those sources of information to which the reader can in fact "refer." Personal communications and unpublished data are not referable sources; they should not be included in the reference list. If you have given a reference as "to be published," be certain that that is the case in fact; if it has not been accepted for publication you do not know for a fact that it is "to be published." Most journals rightly do not accept "submitted for publication" as a reference because the likelihood of publication is still questionable. "In press" has become a standard form for referring to an article that has been accepted for publication but has not yet been published. That status should never be presumed; if you know for certain that the article has been accepted and will be published, use that form; but, if you have any doubt, consider it "unpublished data." Information that you want to cite that has *not been published* should be included in the text itself, with appropriate parenthetical or footnoted information given, e.g., (personal information from Donald J. Duck, L.L.B., in his unpublished report on quackery).

A remarkable effect in the conversion of a researcher into a writer is that he tends to want to use all the resources he has come to know and love throughout his many moons. Perhaps he believes a long list of journal articles and books attributes to him the quality of "having knowledge" about the subject matter. In fact, to the contrary, the truly incisive mind can cut through the thick to the thin.

Journals today abhor long lists of references. You do not have to account for each mot and mite in your paper with a definitive seminal article. In fact, you can ignore references that merely support universally accepted facts. Be wary of extending your list beyond ten references when your message is comparatively short, and beyond twenty when it is of "standard" length. Occasionally authors are permitted many references, but only because of special circumstances, e.g., a "special" anniversary article or a review by a singular expert in a specific area.

Do not, of course, include references when (1) they are not strictly pertinent to the topic, (2) they merely provide incidental information to the point of the message, (3) they are given merely because the authors in the reference are famous and their activities are related (tangentially) to the message, or (4) you have not read them.

A further point to emphasize about references is that they should be accurate to the minutest accent and middle initial. Journal style will determine the form of the references, but responsibility for accuracy of authors' surnames and initials, specific words in titles and subtitles, names of publishers and publications—*all* details—will rebound to you, the writer. Do not rely on a secondary source for such details, i.e., indexes or articles quoted in other articles.

Systems for presenting references, footnotes, and bibliographies differ from journal to journal, usually depending on the general style acknowledged within a given discipline. Thus, the fields of psychiatry and psychology tend to follow one style, physics and other "basic" sciences another, medicine a third, and so on. Even so, styles vary within a single field and an author must refer to the journal he has selected for its particular and perhaps peculiar style.

TABLES AND FIGURES

Many journal editors suspect that readers prefer to jump into an article's tables and figures before even reading the text. Tables and figures are encouraged when they supplant great lengths of cumbersome text; nonetheless, the space that can be given to tables and figures is limited. Most publications are quick to specify guidelines for consolidating data into tables and figures; the number to be included with an article; and the consequences of trying to squeeze in too much information or too many figures and tables—increased costs (you may be asked to share or bear), technical problems (in having to print the giant tables as special inserts at considerable, regrettable, and unacceptable costs), and the outcry of others whose tables and figures are "equally as impressive."

Again, whether in tables, figures, or text, tell your tale briefly.

IN THE WAKE

Having been brought under the glass (and being there still), IMRAD survives, but not without scrutiny and concern. That IMRAD sprang from the logic of the scientific method does not excuse its being used in the preparation of materials that de-fuse and make diffuse the gist of given messages. Rejection and referee systems aside, the writer whose theme and message do *not* suit the configuration of the "research report" (question, design, methods, answers) ought not try too hard to shape it as such.

Increasingly, journals are becoming aware of the scientist's need for more than IMRAD information, and they are giving space and deference to news, commentaries, and poignant insights into profound universal truths. Truth, in whatever form, is finding its way onto the pages despite the prescriptiveness of form and style.

How much better served would the readership be if the article were organized in terms of the priority of value of the information given? That way, the reader would begin with the important things and continue reading until he had come to a place where he would not need to continue further; beyond that point would be materials that he was familiar with and would be gristle, not grist, for his mental mill.

Thus, the Results could be given first, leading into, if not commingled with, points of Discussion. The Introduction would be absorbed into the Discussion, which itself could then become the major part of the article. Materials and Methods would be placed last in the sequence, to be read only if of interest to the particular reader. The new formula would be ARDCaM (*A*bstract/*R*esults/*D*iscussion/*C*onclusions/*M*ethods). Other combinations would be possible, each used to suit the objective or need of a given article. Perhaps the future will see such article-organization acronyms as ADCRaM, ACDRaM, ARCDaM, and AiRaDe (reintroducing a small *i*ntroduction and following the *D*iscussion with a pithy *e*nding).

The drama is change, and vice versa.

16. The Skeleton Key

BEFORE YOU BEGET IT

So that you will be less likely to regret it, the following guidelines are for you to consider before you begin writing, or after you have written a first draft.

The form your article will take should be influenced by the journal or publication you have hopefully selected to be the publisher of your article and the instrument of your fame. We suggest that you do "market research" on the journals you would consider, following these three steps and answering the implied questions.

Step 1: Decide on the audience you expect to read your article.

- Specialists
- Subspecialists
- Generalists in the science
- Scientists generally
- Scientists and some related professions
- Technicians
- Lay public

Step 2: Pick your journal.

- Consider several that may be appropriate for your article.
- Carefully review each for:
  - –Content (Does it seem to match what you have to offer?)
  - –Audience appeal (the same audience as your target audience?)
  - –Advertising (Do the advertised products or services seem to relate to your topic in any way?)
  - –Use of figures and tables
  - –The average length of the articles
  - –Publication latency (How long from manuscript receipt to publication?)

Step 3: Carefully review the "Information for Authors" page from the journal you have chosen.

- Note any particular details pertinent to your article: color, oversized tables, multiple figures, style of math formulas, etc.

OUTLINE AS YOU BEGET IT

Although some writers can dictate an entire article without the benefit of a definitive outline and, with but a few minor revisions to the transcripted manuscript, see it published early in the *Journal of Senior Circulation*, most writers find outlines helpful at some drafty stage. In fact, outlines are helpful in many ways:

- In organizing and writing the first draft
- In preparing the abstract or summary
- In preparing spin-off works such as professional lectures, class notes, and even letters to colleagues and friends.

An outline is by no means required before or while you write your article. Indeed, some authors are stymied and stifled by outlines. However, it is a good idea to try working with an outline at least once to find out if the mode suits you. Because of the nature of the scientific approach to experimentation, many scientists find writing outlines facile and use them as excellent organizing tools. For them, the outline becomes a type of laboratory protocol or even a little pilot study. Actually, an entire outline can easily be evolved during the course of an experiment–with the hypothesis stated at the beginning of the research project, the steps of the method briefly described, and results listed as data are analyzed.

Choose the Form

When your research is a *fait accompli*, develop the outline with the journal you have selected in mind and choose an appropriate form of article for that journal: IMRAD? straight commentary? case report?

Outline in Line

Assume, for illustration's sake, that IMRAD is appropriate for displaying your research in your selected journal. The first marks you should make on paper are rather skeletal and simple: just write the name of each formal section of the IMRAD used by the journal. To leave plenty of space between each heading, write each heading on its own clean sheet of paper. Assign a Roman numeral to those headings.

I. Introduction
II. Method and Materials
III. Results
IV. Discussion
V. Summary/Conclusion/Abstract

That was simple, but you are not yet finished with this step. Each of those headings represents a sizable number of words, some more than others, but each sizable enough to warrant the assumption that there is a succinct theme to each, that is, a major idea you are expressing. Everything you will say in a given section should relate to that major idea. Formulate the theme of each section in a single sentence and write it after the appropriate heading in the outline.

I. Introduction: . . . (thematic statement)
II. Method and materials: . . . (name method)
III. . . . etc.

If you have already done some scribbling and scratching, the next activity will be easier than if you are thinking about your article for the first time. In neither circumstance will this step be simple. It will require thought and concentration. It will bring you rat-to-data with the issues of your article.

From the chapter on paragraphs, you know what a topic sentence is and how it functions in a paragraph. How are you at writing a dozen or so topic sentences that so perceptively identify the important ideas of your article that by listing them you can follow straight through your article with pristine logic? No matter, try it anyway.

Take your outline, which at this stage has Roman-numeraled headings with thematic statements, and, section by section, reflect on each thematic statement. On scratch paper, scribble down the main points you think should be made about that theme. Remember, these are only main ideas, not the supporting details or the supporting reasoning that prove or illustrate the ideas. These main ideas are the topics of the section's theme and, as such, warrant a full paragraph (long or short) of explanation. The one-sentence statement of the topic is the topic sentence of the paragraph, and in the outline each topic sentence should be assigned a capital letter, beginning the list with "A."

I. Introduction: (thematic statement)
  A. (Topic statement)
  B. (Topic statement)
II. Method: (state method)
  A. (Description of patients or materials)
  B. (Description of method in study)
III. Results:
  A. (Topic sentence–results, category 1)
  B. (Topic sentence–results, category 2)
  C.
IV. Discussion
  A. (Conclusion 1)
  B. (Conclusion 2)
  C. (Implications)

If you have done that much, that is, if you have thought through all the main ideas in topic form, you have done a lot. In fact, the crisis is over and what follows is the fun of developing each topic by dissecting it into its various components. Those components may be dissected by logic (inductive or deductive), by cause-and-effect relationships, by physical description, by chronological directions, or by syntheses. Each component of the topic, expressed in a full sentence, should be assigned an arabic numeral, as in the following outline:

Relevancy of Audit*

I. Introduction: Is a formal audit of professional services rendered in a teaching situation useful?
  A. With an emphasis on cost control, PSRO audits medical services paid for by SSA but not those provided in teaching hospitals.
  B. An experiment of house staff in a teaching hospital tested the usefulness of medical audit.
II. Method: Comparison was made of three wards, one designated experimental, during twenty-four months.

*Developed from Peter E. Schrag and William A. Bauman, "Is audit relevant to the medical wards of a teaching hospital?" *Arch. Intern. Med.* **136**:77–80 (January 1976). © 1976, American Medical Association. (See Figure 23-1 for the article reproduced in full.)

A. Three attending physicians were responsible for the experimental ward.
    1. Each assigned an intern.
    2. All worked with resident.
    3. Salaries were pooled and shared.

B. Patient admissions alternated.
    1. Patients were informed.
    2. Patients were billed.

C. Four factors were measured:
    1. Nature of problems.
    2. Length of patient stay.
    3. Average hospital charge per patient.
    4. Number of major diagnostic procedures.

D. Data were collected and analyzed from:
    1. Hospital's computerized billing system.
    2. Medical records.
    3. Self-reporting of attending physicians.
    4. Questionnaire of house staff members filled out anonymously.

III. Results: The results were unexpected.

A. The audit of cost and length of stay showed no important differences.

B. Attending physicians' intervention was helpful in six aspects.
    1. Unnecessary invasive diagnostic procedures were discouraged.
    2. Preparation for continuing care of long-term effects of chronic illness was accelerated.
    3. Dying patient was protected from fruitless intensification of care.
    4. Consultations with other services were expedited.
    5. In emergency situations, senior help was more available.
    6. Decisions about patient management that were made late in course were influenced.

IV. Comment

A. The lack of differences in the four factors may mean personalized supervision is superfluous for those matters.

B. Perhaps the audit was not sufficiently sensitive.

Whether or not you develop your outline further in even more detail is a matter of personal style. Some folks like to itemize so thoroughly in outline form that they need only supply transitional and connecting phrases to actually write the article. Others prefer to let thoughts flow more freely, more randomly, on paper at the outline stage. Some even find it best not to proceed so far with an outline and perhaps make their statements in telegraphic phrases rather than sentences, producing what is called a topic outline.

Whichever, when you have completed the outline, study it for overall flow of thesis. Does each topic sentence lead in some order, whether logical, chronological, or spatial, to the next topic sentence? If not, you probably have some rearranging to do, or maybe you have left out an important thought-step. If two topic sentences within a single section seem redundant, maybe combining them is in order.

You probably remember the rule that if there is an "A" there must be a "B," if a "1," a "2." The reason is that the divisions of an outline divide thoughts into components. With only one component, only one idea is given with no division. The component should be represented, therefore, as a single idea. Conceivably, for example, the Introduction might in one, short, succinct paragraph state the central thesis of the entire article. In that event, the sentence accompanying Roman numeral "I" should suffice. For example, the following single-paragraph Introduction introduces the problem and the approach that could be summarized in one sentence.

Example:

I. INTRODUCTION: In a prospective and integrated study, we measured numerous physiologic factors in uremic patients.

The broad spectrum of physiologic abnormalities producing hypertension involves a quantitative interrelationship of factors that include cardiac output, peripheral resistance, intravascular volume, extracellular fluid volume, and the renin-angiotensin system. Unfortunately, most of the studies of uremic hypertension have considered these factors separately, and a comprehensive integrated study in a homogeneous group of patients is lacking. During the past three years, we have undertaken a prospective study of uremic patients prior to entering a hemodialysis program. In this study, simultaneous measurements of cardiac hemodynamics, intravascular and extracellular fluid volumes and

> the state of the renin-angiotensin system have been made. In addition, we have also studied the response to lowering of blood pressure in the hypertensive uremic patient.*

However, if the Introduction has at least two components, these should be specified.

Example:

I. INTRODUCTION: Computer-averaging techniques permit more sensitive recordings of nerve activity.
   A. The method has been used to record peripheral nerve activity by applying electrodes to the skin.
   B. Our study compares responses at skin level with those recorded directly from the surface of the spinal cord.

> With the introduction of signal-averaging techniques (Dawson 1947; Brazier 1960), it has become feasible to record small electrical potentials. This method has already been used to record peripheral nerve activity by applying electrodes to the skin (Liberson et al. 1966; Kline et al. 1973). Liberson et al. (1966) recorded the transmission of evoked activity successively through peripheral nerve, spinal cord, and cerebral cortex using these techniques. Cracco (1973) has also recently reported that peripheral nerve stimulation evoked potentials in the spinal cord which could be recorded from the overlying skin. He stimulated hindlimb nerves and recorded evoked potentials from lumbar, thoracic and cervical regions. The relationship between these skin level recordings and direct recordings of spinal cord activity has not been experimentally established.
>
> This study was undertaken to compare responses recorded directly from the surface of the spinal cord with those recorded at the skin level overlying the cord. Computer averaging techniques were used for both. The increased sensitivity of computer methodology permitted recordings of conduction over greater distances of cord than in previous studies.**

*Jose L. Cangiano et al., "Normal renin uremic hypertension: Study of cardiac hemodynamics, plasma volume, extracellular fluid volume, and the renin angiotensin system," *Arch. Intern. Med.* **136**:17-23 (Jan. 1976).

Leo Happel et al., "Spinal cord potentials evoked by peripheral nerve stimulation," *Electroencephal. Clin. Neurophysiol.* **38:349-354 (1975).

But really, the proper form is not so important as the flow of development, and you should not let such considerations block your writing process so early in the article's development. At this stage, just get your thoughts in order and under control. As you write the sections, you can refer back to the outline, magically altering its structure with your acquired flashes of perception that beam forth from the illuminating writing process itself. To make your outline even more useful, you might leave an oversized margin on the left or right, in which you can make notes concerning figures, tables, and references corresponding to appropriate topics.

When you have finished the penultimate draft, you will file this outline and take a different approach to outlining what you have written—an editorial approach.

17. Physiology of the Article

READING PROCESS

As physiology deals with the functions and vital processes of living organisms, we use the term here metaphorically to mean the dynamic processes that occur in and among the parts of the article. Such dynamics evolve from principles of communication, learning, and memory, and you become involved in those processes by writing an article—like it or not.

We now consider real-world methods used by readers for getting information from published articles and the ploys used by publishers to satisfy reader needs and facilitate information-getting. As described in a preceding chapter (15), the form of the science article has lost some of its absoluteness in the past generation and many publishers appreciate that readers prefer "modern" methods for getting information from the contents, but IMRAD still dominates the literature.

To Read or Not To Read

If you have time, you may want to turn each page of the journal, thus discovering some of the givens in the contents—new equipment advertised, new drugs described, interesting filler materials not listed in the table of contents, and special or regular materials. If, however, your time is in demand, you will need efficient methods for reviewing the contents of journals, articles, and even parts of articles. Some guidelines have been published on how to improve efficiency in "Keeping up with the Journals."*

The importance of the interrelationship of the parts of the article is evident in the way the reader looks at articles today. Clarity and rapid comprehension are achieved with expected form. Expected form is the progression from one unit to another according to an

*David L. Schmidt et al., "Keeping up with the journals," *Rocky Mountain Med. J.*, pp. 51–54 (January 1973).

orderliness that the reader anticipates. In science and medical writing, we have come to depend largely on the orderliness of IMRAD.

A typical approach to reading science literature is to glance at the titles in the Table of Contents of a journal. A title is a go or stop sign, depending on your interests; key words there will help you decide whether the article is worth further time. If *go*, then flip to the first page of the article and read the Abstract, which essentially says, "This is what the problem was, what was done, and what was found out." If at this point the details interest you, proceed with your reading. Two forms–the title and the Abstract–will have facilitated your deciding whether or not to become involved with the information in the article. If, on the other hand, after reading the Abstract, you perceive that you are not interested in reading the article, the forms have saved you considerable time, enabling you in a few minutes to judge the article's value to you rather than spending many more minutes reading the article, and sparing you the existential struggle of deciding whether to read or not to read, only to learn you should not have bothered.

Clue the Reader In

The form of the remaining sections of the article are even more expected than you might have expected. If you just extract the headings and note the movement of concepts they lead you in, you will see that the format really follows the scientific method. As writer, as researcher, did you not begin your experiment with a problem, a question that you asked–what did you want to find out? Can you state the very question that your mind was ruminating when you formulated or first encountered your problem? In other words, what was your hypothesis? A statement of that hypothesis is the basis of the Introduction of the article. If you pose that question to the reader in the beginning, the article becomes a kind of mystery story and he is invited to become involved, just as you did, in the quest for the answer to that question. Maybe you had some clues from the beginning; if so, you should reveal these clues to the reader in the Introduction so that you start out "with one mind" to solve the mystery.

Having asked the question, you had to set up an avenue of investigation–how do you further investigate those clues? Can you set up a situation that reenacts the reality you want to know about? You devised a way to investigate the question in your laboratory (be that the room down the hall with machinery and test tubes, the hospital clinic, an entire town, or a river). Your method of getting at the answer became definite and systematic as you described it in a protocol for yourself and your co-workers. Now describe it so the reader can follow the steps in recreating the reality and, if he chooses, he can create that reality in a laboratory himself. Call this description your Methods section.

Once you set your experiment in motion, things began to happen. As they happened, you noted whatever occurred–more clues. Reveal these to your reader; they are the Results section of your article.

Finally, when the laboratory study had gone far enough, and you had all the clues possible, you sat down with the clues, walked with them, nightmared, struggled, and wrestled with them, talked about them and maybe to them, and finally came up with some concrete thoughts about them. You came up with some answers, maybe only partial answers, or maybe you found out what dunnit and solved the mystery. Whichever, the reader waits to hear your conclusions, and if the underlying reasoning is complicated, he also wants to know that reasoning. Meanwhile, he probably has been guessing about the conclusions himself, but the short time it takes to read an article is not time enough to consider all the implications. So share your thoughts as succinctly as you can in the Discussion and Conclusions sections of the article.

Dosage 55: Expected form

(1) Read Ben Masselink's article "Everything You Wanted to Know About Television But Were Afraid to Ask" (*Playboy Magazine*, July 1970).

How is the title "expected form"?

How is the development of the article expected form?

How do those forms tell you what the real subject under discussion is and lead you to expect certain themes?

(2) How are genetics expected form?

Genetics imprint. Ducks imprint within the first two days.

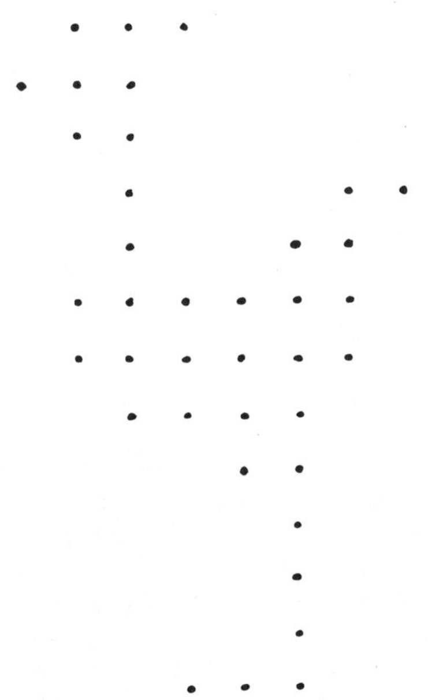

Fig. 17-1. Dots a duck, Clyde.

Can you write an article entitled "Everything You Wanted to Know About Genetics but Were Afraid to Investigate?"

(3) Read the following letter.* Does any of it sound familiar? What is your first clue the letter is not really to Mr. Jefferson?

July 20, 1776

Mr. Thomas Jefferson
Continental Congress
Independence Hall
Philadelphia, Pa.

Dear Mr. Jefferson:

We have read your "Declaration of Independence" with great interest. Certainly, it represents a considerable undertaking, and many of your statements do merit serious consideration. Unfortunately, the Declaration as a whole

*Edward Schwartz, *Social Policy*, Vol. 5, no. 2, July/August 1974, pp. 10-11.

fails to meet recently adopted specifications for proposals to the Crown, so we must return the document to you for further refinement. The questions which follow might assist you in your process of revision.

1. In your opening paragraph you use the phrase "the Laws of Nature and Nature's God." What are these laws? In what way are they the criteria on which you base your central arguments? Please document with citations from the recent literature.

2. In the same paragraph you refer to the "opinions of mankind." Whose polling data are you using? Without specific evidence, it seems to us, the "opinions of mankind" are a matter of opinion.

3. You hold certain truths to be "self-evident." Could you please elaborate. If they are as evident as you claim, then it should not be difficult for you to locate the appropriate supporting statistics.

4. "Life, liberty, and the pursuit of happiness" seem to be the goals of your proposal. These are not measurable goals. If you were to say that "among these is the ability to sustain an average life expectancy in six of the 13 colonies of at least 55 years, and to enable all newspapers in the colonies to print news without outside interference, and to raise the average income of the colonists by 10 percent in the next 10 years," these would be measurable goals. Please clarify.

5. You state that "whenever any Form of Government becomes destructive of these ends, it is the Right of the People to alter or to abolish it, and to institute a new Government. . . ." Have you weighed this assertion against all the alternatives? Or is it predicated solely on the baser instincts?

6. Your description of the existing situation is quite extensive. Such a long list of grievances should precede the statement of goals, not follow it.

7. Your strategy for achieving your goal is not developed at all. You state that the colonies "ought to be Free and Independent States," and that they are "Absolved from All Allegiance to the British Crown." Who or what must change to achieve this objective? In what way must they change? What resistance must you overcome to achieve the change? What specific steps will you take to overcome the resistance? How long will it take? We have found that a little foresight in these areas helps to prevent careless errors later on.

8. Who among the list of signatories will be responsible for implementing your strategy? Who conceived it? Who provided the theoretical research? Who will constitute the advisory committee? Please submit an organizational chart.

9. You must include an evaluation design. We have been requiring this since Queen Anne's War.

10. What impact will your program have? Your failure to include any assessment of this inspires little confidence in the long-range prospects of your undertaking.

11. Please submit a PERT diagram, an activity chart, and an itemized budget.

We hope that these comments prove useful in revising your 'Declaration of Independence."

Best Wishes,

Lord North

Dosage 56: Other forms

Read an article in *Science* magazine or *New Scientist*. How do these depart from expected form? Do the departures help or hinder your comprehension and anticipation of the ideas?

Improving Your Process

If you are too far behind in your reading and believe you might be a slow reader, you might find helpful some of the articles that have been published to improve your speed and comprehension [e.g., E. I. Vernon and I. T. Horrocks, "Speed and comprehension in reading," *Med. Opin. Rev.* **6**:52–55 (Nov. 1970)].

READER DYNAMICS

Physiology extends beyond the mere content of the article and involves communications theory and practice, pattern and form, conditioning and learning, and a great unlimited number of implications. How can one read the word "detente" without automatically thinking of Henry Kissinger and his foreign policy and of Gerald Ford and his efforts to circumvent use of the word altogether? In writing, one is supposed to be able to educate, persuade, and occasionally surprise and entertain. These are qualities that extend beyond the mere givens of words strung out to form specific images; they play with the mystical, inexplicable, and enigmatic abstractions called creativity, intellectual excitement, humor, "reading between the lines," irony, and paradox.

The effectiveness of the process largely depends on the cultural and educational background of the reader himself. To read, "We call that chicken Tom, because he's always 'peeping'" involves the pun of "peeping Tom" and the chicken who peeps. Without an understanding of the two interpretations, the derivative humor is lost to the reader, and the sense of what is written is left to show merely the naming of a chicken. If you are writing for a technical or specialty journal, you can assume that most of your readers are of an educational level to understand the content of your article, but their cultural backgrounds may vary widely—no doubt foreigners will read the journal. Also remember that people in other disciplines may read your article.

The overall effect on the reader as he looks at your article, then, should coincide with your own overall objective in writing the article in the first place. If you are showing a method for solving a problem, you must consider which approach to use in explaining that method. Often a reader will not get the "insight" needed unless the problem is approached from an unexpected perspective, one that suddenly shifts the viewpoint so that he can cross the perceptual threshold with a new perspective.

A simple game problem can be used here as a Dosage.

Dosage 57: A dot-of-view

Connect the nine dots in Fig. 17-2 by drawing four continuous straight lines without raising your pencil from the page.

Unless you already know the solution, you will be inclined to limit your trials and errors to the area prescribed by the nine dots. If, however, you are given a new insight, a shift in the instructions, that

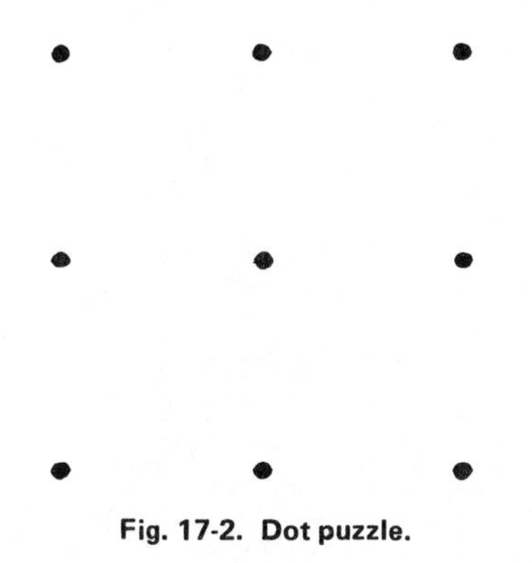

Fig. 17-2. Dot puzzle.

you can go beyond the area delimited by the nine dots, then the solution in drawing the four continuous lines becomes much easier.

Within the structure of your article, then, you must decide at which point to tell the reader what he should know about the stuff of which the article is written, and also whether the reader needs to be told more than once about a given point. The guideline used by publishers that "Half as long is twice as good," ought *not* to force you to omit information that is essential to the proper impact and interpretation of the article.

Reader Attention and Motivation

In the context of your writing, you must always consider how to motivate the reader to continue reading. Presumably your message is important, and its importance would be lost for the reader were he to skip vital parts of that message. Surprisingly, simplicity is an effective means of getting the reader's attention—simplicity in stating what the objective of your article is, what questions it will raise and answer, and what the reader should learn from it. Unfortunately, the complexities of science cause science articles to deal with complicated concepts and relationships. Nevertheless, each new concept and relationship can be presented with consideration for the audience, simply enough to maintain interest and attention.

The reader's own perspective on the subject of your article may be one of "cognitive dissonance," whereby he would consider some

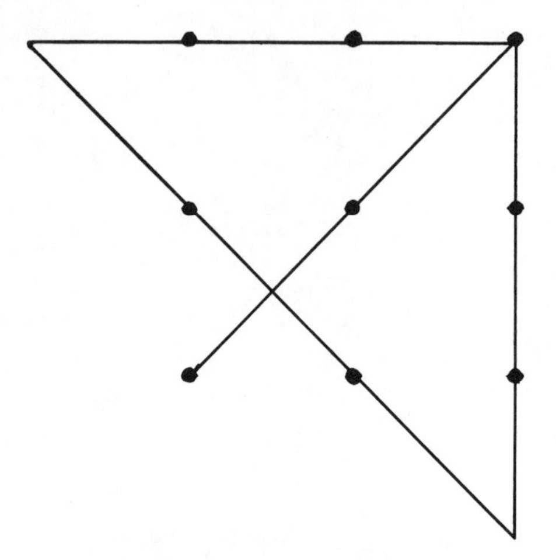

Fig. 17-3. Dot puzzle solution.

things related in ways that they are not, in fact, related. If you must persuade him to a correct viewpoint, the simple approach may be the best. Some would argue that persuasion is *not* the intent of science writing, but how then would they account for learning? Learning is measured by change, change in attitude and in behavior. Persuasion, by whatever name, is implicit to learning. And the preceding paragraph is itself an example of simply stating something as a means of persuading the reader.

In and Out of Mind

The writer presents to the reader information that should be useful to the reader in some way, if only he can learn it—that is, receive it, store it, retrieve it, and then apply it. The writer has a positive force attracting his information in the form of the reader's urge to know, his curiosity, his drive to resolve uncertainty. Unless the writer gives the reader encumbrances that frustrate that urge to know, the reasoning presented should impress the reader's mem-brain so that it later is directly retrievable for forming new concepts. That retrieval system is, very simply, *memory*.

Confounding memory is the mental quality called *forgetting*. In teaching us about memory, researchers have taught us about forgetting as well. Forgetting is an essential function in ridding the mind's storage of irrelevant and useless information. (In what key did your gut rumble before lunch?) One contributor to forgetting, called crowding, pertains to a condition in which too much information is given in a small interval. Another contributor to forgetting is interference, whereby information presented interrupts the reasoning between related points and "interferes with" the reader's comprehension.

A benefit in having others read your manuscript is their feedback on whether you have included too much information, have detached related points, or have not adequately emphasized ideas in appropriate ways. They may not say so in such specific terms as "interference" or "crowding," but you will be able to read between the lines and get that message from their message about your message.

As you discovered through free association (Dosages 12 and 13), the mechanism of information retrieval from the mind apparently can take many channels. The more direct relationships the writer can show for each fact, the more numerous will be the direct pathways of recall for that fact. Use concrete images if possible. Shape, color, size, texture, or position are criteria for presenting such images. The new taxonomy for viruses uses several unique categories, such as these examples.

Capsid symmetry	Reaction to ether
Virion: naked or enveloped	No. of capsomeres
Site of capsid assembly	Diameter of virion (nm)

New Concept? Imagine That!

For the reader to remember a new concept, he will need to "see" concepts in relationship to his own experience or be given some means to form new relationships. Use of mnemonic devices has helped many students remember taxonomies, lists, and such groupings as the twelve cranial nerves (On Old Olympus' Towering Top A Fat-Armed German Viewed A Hop: Olfactory Optic Oculomotor Trochlear Trigeminal Abducens Facial Acoustic Glossopharyngeal Vagus Accessory Hypoglossal). The difficulty with mnemonic devices is that they themselves become things to remember. With such aids as Old Olympus, the memory may require another mnemonic device to recall the correct order of associations—whether the Optic, Olfactory, or Oculomotor comes first.

Helping the reader make visual associations may help him perceive your ideas more effectively. One such help would be to relate the subject to something tangential to any expectation. If you are discussing temperature-sensitive virus mutants, try relating them to

---

Dosage 58: 30? 31?

Can you recall what the mnemonic device illustrated below is supposed to help you remember?

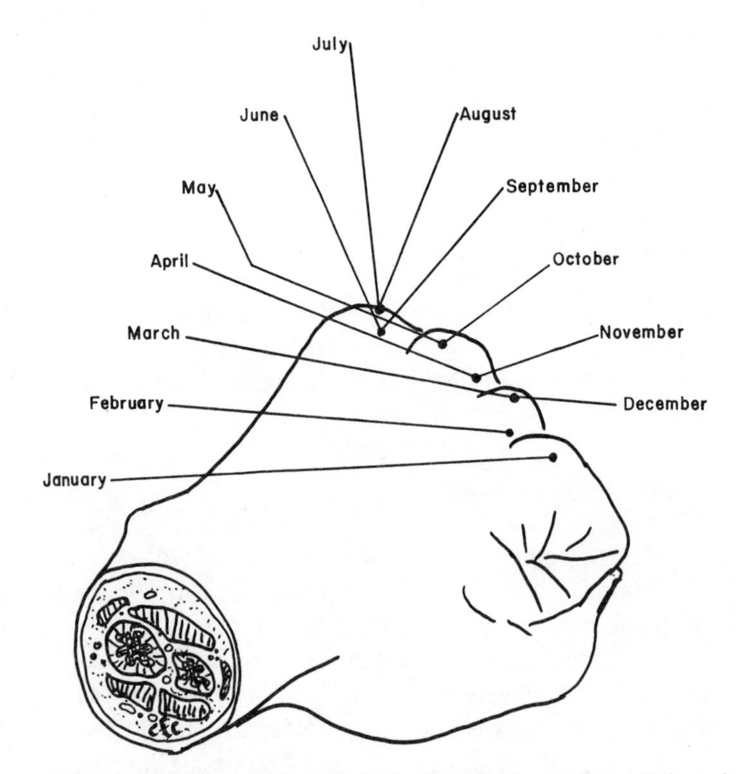

Fig. 17-4. Mnemonic aid for recalling months that have thirty-one days and those that have fewer (thirty or twenty-eight—or sometimes twenty-nine!).

---

music, golf, or planning a garden. Thrown mentally into a strange setting, the reader can be more attentive and ready to absorb new information. Be beholden to Tschaikowski's *1812 Overture*, active, robust, and stirring, in characterizing the mutant *1510* as active, robust, and stirring at one temperature but slow (largo) at another.

The Critical Reader

The "art" of writing is in the style with which the words are put to paper, and therein lies a problem. Often the writer will shade data by the tone and emphasis of his sentences, leading the reader to read as objective what is really subjective, or as subjective what is really objective. The writer should be aware that the "objective evidence of words on paper" will be interpreted specifically as such by each reader of those words.

The ultimate test of objectivity, of course, is whether the statements given as true are in fact known to be true. Sometimes "unsupported assumptions" slide by the critical eyes of reviewers and editors and the innocent but critical reader is left to decide for himself the validity of the assumptions. For example, the statement that "more money is needed to overcome the crime problem" could lead a reader to believe that an authority knows that such is true. The critical reader, however, will ask the important question, "What support does the author give for that statement?" If no support is given, if no data are presented to show circumstances in which more money helped to overcome the crime problem, then the statement is an unsupported assumption that may or may not be true.

A writer who comments that "Most scientists prefer . . ." or "The

---

Dosage 59: The critical reader

For an idea of how one reader might read and interpret an article, read the following two articles:

Ruth D. Abrams, Denial and depression in the terminal cancer patient. *Psychiatr. Q.* **45**:394-404 (1971).
Raul H. Vispo, Critique of 'denial and depression.' *Psychiatr. Q.* **45**:405-409 (1971).

---

concept has been well received . . ." had better be able to give some evidence for such statements. Otherwise, the statement should be couched in qualifying terms.

Visual Effects

Part of the physiology of the article, the process of communication, is the typography and graphics format of the journal. Too little research has been done on whether one typeface is better for transmitting information than another. Some typographic "experts" are inclined to favor modern, streamlined typefaces that many readers find difficult to read. Some journals, for example, use a sans serif* typeface too closely spaced, which is hard to read because it inclines the eye to a vertical movement rather than a horizontal movement. Such a typeface might facilitate a vertical speed reading technique but impedes thoughtful reading.

Dosage 60: Typeface

Read the following samples. The first is in a sans serif face, the second in a serif face. Which do you find easier to read across the page? Ask your typist to type samples of all styles available to you and decide which is best for single-spaced grant proposals and reports and which for double-spaced manuscripts.

One day as Cunegonde was walking near the castle in the little wood known as "the park," she saw Dr. Pangloss in the bushes, giving a lesson in experimental physics to her mother's chambermaid, a very pretty and docile little brunette. Since Lady Cunegonde was deeply interested in the sciences, she breathlessly observed the repeated experiments that were performed before her eyes. She clearly saw the doctor's sufficient reason and the operation of cause and effect.

One day as Cunegonde was walking near the castle in the little wood known as "the park," she saw Dr. Pangloss in the bushes, giving a lesson in experimental physics to her mother's chamber-

*Serifs are short lines angled at the ends of letters. Sans serif letters have no such lines.

maid, a very pretty and docile little brunette. Since Lady Cunegonde was deeply interested in the sciences, she breathlessly observed the repeated experiments that were performed before her eyes. She clearly saw the doctor's sufficient reason and the operation of cause and effect.*

With language being learned by apes and with the advance of reading machines, all the aforesaid may well be considered moot before the present century is spent. And so we come full cycle, back at the grunt with Adamape and Evape.

*From *CANDIDE,* by Voltaire, translated by Lowell Bair. Copyright © 1959. Bantam Books, Inc. By permission of Bantam Books, Inc.

PART V. PRINT AND PRODUCT: "EDITED" UNTO "CREDITED"

18. What Goes, What Ought?

EDIT ONCE YOU BEGET IT

Put the manuscript out of sight and out of mind for a week or more. Actually, the longer you can banish it from view and thought, without losing interest or missing deadlines, the better. The advantage is that when you pick up the manuscript again it will have a freshness—or, heaven forbid, a staleness—as though someone else had written it, and you will be able to edit it more critically. As you reread what you wrote, you should find phrases that need rephrasing, paragraphs that need restructuring, and materials that need discarding. As you try to follow your own reasoning in the article, the gaps in logic and necessary information will become more evident to you. Until you have "detached" yourself from the material for a time, your mind will tend to overlook or fill in gaps, and you will make assumptions about your reader's ability to perceive the meaning. So, for a while, forget it.

This final editing is an extremely important operation in the writing process; no author should submit a work for publication without thoroughly reviewing the manuscript, including the continuity of the material, the logic of its order, the accurate integration of figures and tables, and the accuracy of all information, especially that given in the tables and references.

When you do edit the manuscript, you should use three approaches, suiting your own personality with the way you integrate those approaches. The three approaches may be combined according to the following three patterns.

(A) Whole to Part to Detail
    Look at the sense of the whole article.
    Analyze each section of the article.
    Scrutinize each word for effectiveness.

(B) Part to Whole to Detail
    Analyze the sections of the article for the theme of each.

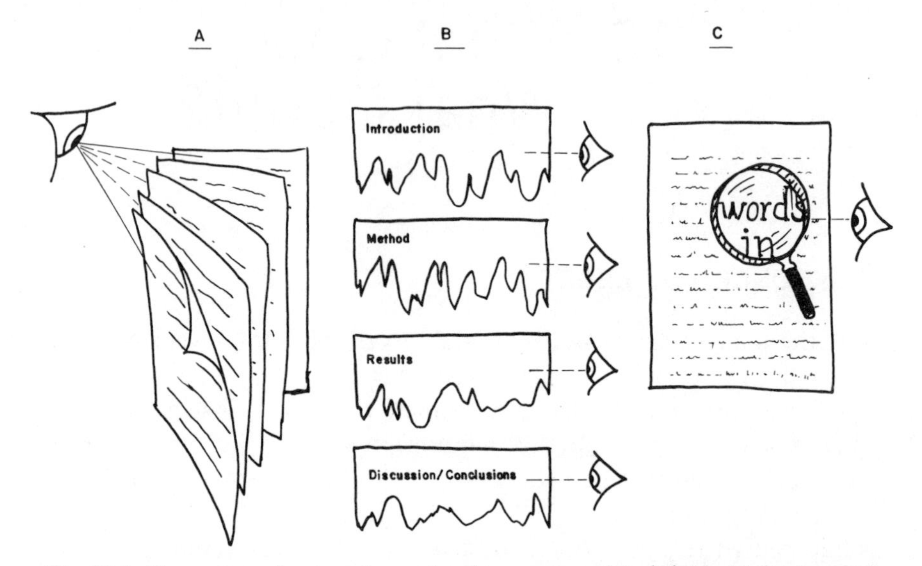

Fig. 18-1. Three approaches to review and analyze a manuscript. (A) Considering the whole thing. (B) Reviewing each section as a subwhole. (C) Carefully scrutinizing each word.

Consider whether they unite to create a meaning for the whole article.
Scrutinize each word for effectiveness.

(C) Detail to Part to Whole
Scrutinize every word for effectiveness.
Analyze the sense of each section.
Consider whether all unites to create a meaning for the whole article.

Each of the combinations has a special appeal for different personalities. The first will no doubt appeal to the deductive reasoners who enjoy thinking about universal truths and absolutes and all things great and beautiful. The second has the advantage of allowing the writer to build rather more inductively, the type of personality that prefers to write each section of the article as the experiment progresses. The third is for those of us who cannot figure out what the whole paragraph or section is trying to say until we understand each word and sentence.

Whatever combination you have chosen as peculiarly suited to your needs, each of the approaches should include the following calculations.

The Overall Message of the Whole

Ask the following questions of yourself and your manuscript:

(1) What, specifically, have I said? (Answer in one sentence.)
(2) Is it worth saying? (Be honest. Does it contribute anything to our knowledge?)

Organization of the Sections

Hide the outline you made way back when you started writing your article. You are going to make a new outline now, not of what you will say but of what you have said. This post facto outline will illustrate what is actually on paper, not what is in your mind. It will be a tool for analyzing the logic, unity, and coherence of each section. This step is a bit like the checking you do for a subtraction entry in your checkbook–for example:

$$\begin{array}{r} \$\ .53 \\ -\ .25 \\ \hline \$\ .28 \end{array} \qquad \begin{array}{r} \$\ .25 \\ +\ .28 \\ \hline \$\ .53 \end{array}$$

To check your subtraction, instead of repeating the subtraction activity, you add the subtrahend and the total to get the minuend.

So, now make a sentence outline of the manuscript, divided according to section headings. Indicate each section by a Roman numeral, and each paragraph within the section by a capital letter.

I. Introduction
  A.
  B.
II. Materials and Methods
  A.
  B.
    1.
    2.
      a.
      b.
III. Results
IV. Discussion
V. Conclusion or Summary (if appropriate)

Each paragraph of your manuscript should have a topic sentence, whether it ultimately occurs at the beginning, middle, or end of the paragraph. Identify the topic sentences and write them into your outline. After you have entered all topic sentences in the outline, you should be able to read straight through the outline and follow the sequence of your reasoning. If the logic does not follow, you have more work to do, perhaps some reorganizing within a section, perhaps the addition or deletion of material. Does each section of the manuscript relate directly to the question and purpose stated in the Introduction?

The outline also provides a handy diagram for analyzing the internal development of each section.

- Ask yourself whether each section contains the elements it should and whether the section achieves its purpose.
- Do all the paragraphs in the section relate to the section?
- Have you inadvertently included some discussion in the section on results? or some data in the section on materials?

(Review Chapters 15 and 17 to refresh your memory about the anatomy and physiology of a science article.)

Clarity and Style: Words, Sentences, Paragraphs

Now take the article apart sentence by sentence, word by word. Precise words and good sentence structure are necessary for the rapid advancement of the ideas throughout the article. Each sentence should be thoroughly understood and clearly stated before you begin analyzing each paragraph for its unity and coherence.

Words

Are the words *concise?* (Do they "cut into" the exact meaning of your idea?)

Are they *economical?* (Have you said whatever with as few words as possible? Have you said anything that is self-evident and hence need not be said at all?)

Are they *precise?* (Avoid vague adjectives, such as "large amount," and give exact measurements.)

Are there any *stacked nouns* or wordy phrases?

Is there any *jargon?*

Can an *Anglo-Saxon* (English) word be used instead of a Latinate word?
Are the words *ambiguous?* (Do they leave any questions about who or what was done, thought, or concluded?)
Are there any *dangling modifiers*? (Beware of present participles and infinitives.)
Do the transition words and conjunctions convey the correct *logical link* between two parts of a sentence or between two sentences?

Sentences

How do the parts within the sentence relate to one another?
Does each verb have a proper *subject*—plural to plural, singular to singular?
Can the passive voice be changed to the *active* voice (to avoid ambiguity and vagueness)?
Are there any series; if so, are the elements in the series expressed in a *parallel structure?*
Are thoughts in clauses juxtaposed in parallel structures?
Do the sentences *vary* in length and style (simple, compound, complex)?

Once the sentences have been corrected and polished, consider each paragraph as a unit.

Paragraphs

Does the paragraph have *unity*, that is, is it about one main topic?
Does one sentence (the topic sentence) state the *topic* of the paragraph?
Is the paragraph *complete*, that is, is enough information given about the topic?
Does the paragraph develop logically?
Do the sentences have transitional words where appropriate?
Is the next paragraph the next logical thought development in the section?

Dosage 61: Begetting editing

Edit the following paragraphs and text for all the characteristics described above in the three approaches.

(1) Depressive symptoms are not infrequently considered to be the result of organic brain disease. Unfortunately, depressive symptoms are not unusual in patients with organic brain disease; therefore, considerable effort should be given to an adequate diagnosis. It is my impression that between 4% to 6% of the elderly over age 65 demonstrate signs and symptoms of chronic brain disease.

(2) Capuchin Monkeys (*C appella*)–Only two adult female monkeys and no adult male monkeys of this species were able to be obtained. Neither of the adult females had aortic lesions. For aortic lesions, all of the 18 juvenile monkey aortas collected in this study were negative. No coronary artery lesions were detected in the two adult or 18 juvenile *C appella* monkeys.*

(3) **Report of a Case**

The patient is a seventeen-year-old male, who was noted to have "blue eyes" as an infant by his parents. The patient suffered several fractures as a child, and X-rays taken at each occasion were consistent with a diagnosis of osteogenesis imperfecta. X-rays taken in 1965, 1966, and 1967 showed the characteristic changes of osteogenesis imperfecta: narrow diaphysis, widening of the metaphyses, of the radius and ulna, and healed fractures.

Because of the history of frequent fractures, the child was told to avoid trauma and it was recommended that he not attend public schools. A pes planus deformity was treated with elevation of the inner border of both heels and soles with rubber and leather supports. The patient sustained a fracture to a metacarpal in February, 1971, which healed without sequealae. This is the last known fracture.

In June, 1974, the patient was seen at the Earl K. Long Hospital dermatology clinic with a six month history of multiple asymptomatic firm, erythematous papules, 3–5 mm in diameter, located at the left lateral aspect of the neck and symmetrically distributed on the anterior-lateral

*W. P. Newman, III, et al., "Naturally occurring arterial lesions in new world primate," *Arch. Pathol.* 98:173 (Sept. 1974). © 1974 by the American Medical Association.

aspects of each elbow. A shave biopsy was done, and an excisional biopsy followed one week later. Local therapy with liquid nitrogen and intralesional triamcinolone 5 mg/cc. was begun after the diagnosis was made. There has been some flattening of the lesions. All laboratory work has been within normal limits: bilirubin (total), = 1.3 mg%; alkaline phosphatase = 75 I.V., LDH = 67 I.U. sodium = 144 mE/potassium = 5.0 mEq/fasting blood sugar = 104 mg% b.u.n. = 15 mg% hemoglobin = 14.9 grams, and white blood count = 8,300. The patient's electrocardiogram was entirely without abnormalities, the chest X-ray showed no unusual changes, and the patient had a normal waking and sleeping state electroencephalogram. The VKRL was non-reactive, and fungus studies have been negative.

A chromosome study was performed. The number in peripheral leukocytes was 46 and analysis of several metaphases revealed no structural abnormality of the chromosomes.

The physical exam revealed no abnormalities other than the presence of light blue sclerae and the previously mentioned skin changes. The patient was noted to be of average height for his age, but was of slender build. There was no hyperelasticity of the skin or articular hyperlaxity. There was no decreased auditory acuity noted, but an audiogram was not performed.

The patient has been in good health since he first was seen as the hospital. The patient has continued to wear corrective shoes for his pes planus deformity. The patient now attends public schools and leads a rather normal life. His grades in school have been good. The patient is one of four children. His two brothers and one sister are in good health and show no evidence of being affected with either of the diseases under discussion. The patient's mother and father are alive and in good health, and are not affected with EPS or osteogenesis imperfecta.

After you have done as thorough an editing of your own paper as you know how, let your colleagues read what you have written. Ask them for criticisms, suggestions, comments. You may ask them to be ruthless but gentle. Most will be flattered by your request and often will find the "horrible error" that would have been embarrassing if allowed to go to the eyes of a journal reviewer. Even ask friends not in your particular discipline to read it and comment on it. Sometimes those who are too close to the specialty are too quick to presume that some things need not be questioned. Use everyone who could help put the article into its best form. A friend may be an English expert; have him or her pick at the sentences to be sure that all language conventions are properly used.

Under no circumstances consider the manuscript ready for submission until one or more colleagues have read it and you yourself have thoroughly reviewed it.

IN GOOD COMPANY

An experiment was conducted in the redoubtable journal, the *New England Journal of Medicine.* A highly technical article on immunology was published in the usual format and style of technical articles; also published in the same issue was a version of the same article rewritten by a lay science writer. The editor of the *New England Journal of Medicine* solicited comments on the articles from readers.

Dosage 62: I would say . . .

Write out your opinion about the following two excerpts from those two articles. You might want to try your hand at revising the entire article by Gutterman before reading that by Culliton.*

Excerpt from Gutterman

During the past several years there has been increasing evidence that tumor-associated antigens are present in a variety of human neoplasms, including acute leukemia. In studying the immunologic response to human leukemia, Yoshida and Imai found that the serum of 71 per cent of patients with acute leukemia gave a positive immune adherence reaction

*J. U. Gutterman *et al.*, "Immunoglobulin on tumor cells and tumor-induced lymphocyte blastogenesis in human acute leukemia," *New England J. Med.* 288:169–173 (1973); B. J. Culliton, revision, pp. 173–175. Printed by permission.

with autologous leukemia cells. In contrast, using similar methods, Doré et al. observed that only 22 per cent of 140 patients with acute leukemia had evidence of antibody directed against autologous leukemia cells. Cell-mediated immunity was observed by Oren and Herberman, who reported that the majority of patients with acute leukemia had positive skin tests to membrane preparations of their own leukemia cells. Cellular reactivity to leukemia has also been investigated in vitro by the mixed-leukocyte-culture technic and by lymphocyte cytotoxicity. Both in vitro lymphocyte blastogenic responsiveness and lymphocyte cytotoxic reactions to autologous leukemia cells have been demonstrated in patients with acute leukemia.

Of possible relevance to the apparent failure of this tumor immunity to control clinical cancer was the demonstration that serum factors in patients with solid tumors may interfere with in vitro cell-mediated immunity. The Hellströms and their co-workers have described a factor (presumably antitumor antibody) in the serum of patients with solid tumors that may block lymphocyte-mediated cytoxicity to tumor cells.

Because there have been few studies of the role of serum factors in tumor immunity in patients with acute leukemia, the current study was designed to determine the presence and possible inter-relation of humoral and cell-associated tumor immunity in acute leukemia. Evidence for both humoral and cell-mediated tumor immunity was found. The presence of a vigorous lymphocyte blastogenic response to leukemia cells, its inhibition by autologous serum and the presence of immunoglobulin bound to the leukemia cells were observed to correlate with a good prognosis.

Excerpt of Revision by Barbara J. Culliton

The immune system appears to play an important role in determining how a patient will respond to a tumor. Presumably, he has a better chance of survival or, at least, of experiencing remission, if his immune system is alert to the presence of tumor cells. We are not yet certain, however, of the precise mechanism or mechanisms involved in natural tumor immunity. Nor do we fully understand the facets of immune responsiveness that may vary specifically from one type of tumor to another.

During the last several years, there has been increasing evidence that tumor-associated antigens are present in a variety of human neoplasms, including acute leukemia. Concurrently, emphasis has been placed on the cell-mediated lymphocyte response to these antigen-bearing tumors as a measure of immune reactivity. Previously, we showed a correlation between a patient's prognosis and his general level of immunocompetence. In this study, we are extending that correlation to more specific areas of immunocompetence. And we are suggesting that emphasis should be placed not only on cell-mediated, but also on humoral, or circulating response in achieving some measure of natural tumor immunity, particularly with regard to acute leukemia.

Thus, to acquire a more complete understanding of a cancer patient's immune response to his own tumor, it is important to consider both the cell-mediated and the humoral aspects of tumor immunity. Our study of 34 adults with acute leukemia was designed to determine the interrelation of a lymphocyte blastogenic response to leukemia cells (a response that indicates lymphocyte recognition of tumor antigen), any effects the patient's own serum has on this response, and the presence or absence of immunoglobulin on the surface of the tumor cells.

The responses to the articles were a mixed bag. Many in the scientific community objected to the revised article, saying they could not understand it as well as the original and that they found its terminology and general organization difficult. Others praised the effort and the result, finding that the revision improved their understanding of the subject in a way the original article had not done.

What does such an experiment show? One inference is that many readers of journals have become conditioned to the traditional style of the science article and rely on that style to pigeonhole their reading and information seeking. A second inference is that other readers would welcome a change in the style; they, perhaps, appreciate information best when presented in more direct language.

19. Print Put

By the time you arrive at this chapter, you no doubt have a concept of yourself as a writer founded on the accumulation of skills that pertain to writing and on the confidence that derives from success with words, sentences, paragraphs, and the effective representation of ideas on paper.

No doubt, too, the triumph of having prepared a manuscript is sweet on the mind, and the hopes and expectations carry you to spheres labeled "and now" or "presuming that" or "what if." Before you reach the orbit of "Who am I to doubt that this is truly worthy and sufficient to the prize?", you should reflect on certain truths to temper those thoughts, hopes, expectations, and presumptions.

One question that should be asked—that will be asked—is, "Is this manuscript really necessary?" You may see what you have written as the unique answer to a great need in your specific area of interest, but others will not have your parochial perspective. Editors will be weighing your manuscript on a scale of values: mankind *needs* this message, some good general information here, the specialist will find something in the contents, maybe this will help corroborate Dr. Rinky Dinkey's data, maybe this will stimulate someone to go and do a *good* study on this question.

Assuming you have decided your manuscript is necessary, we now enter the mechanical phase, in which we consider the particulars of how to systematically organize a manuscript for submission to a publisher and of what action to take when a manuscript is rejected—beyond declaring the idiocy of the rejecting publisher. Much of the frenzy that writers experience in getting articles accepted and published can be prevented by informed anticipation of the requirements for proper preparation of materials.

The old saw about first impressions being the most important holds for articles submitted to editors for review. If an editor, and later a reviewer, sees that the copy is replete with typos, the form is sloppy, no abstract is given when one is required for that journal, and the references do not follow the form or order that journal uses, the finding of faults in the actual message of the article becomes an easy matter. Skepticism and criticism already abound, appropriately, in a review of the content of a scientific article, so do not spur their geometric increase through the trivia of form, margins, and the misgivings of references.

PREPARING THE TYPESCRIPT

Look before you leap into preparing the final draft. Study the journal you have selected to submit your manuscript to. In at least one issue of each volume, most journals provide a page or two of "Instructions to Authors" or "Information for Authors." Find those instructions. Read them carefully. Then study the form of articles in the journal. If the journal style uses capitals for the first letters of the major words in headings, type your headings with caps and lowercase letters. If the by-line does not show academic degrees, do not include them in your own by-line. Does the journal number each figure and table? Wellsir, then do ye likewise. If each article has an abstract, prepare an abstract for your article. Does the journal use footnotes? What is its style for references? Are lines used in tables? How are acknowledgments worded and where placed? Are numerals used for numbers below ten? Are abbreviations allowed?

Find a competent typist (if you are not one yourself) and have him or her carefully read the "Information for Authors" page. Unless the journal's instructions otherwise specify, observe the following guidelines. The typescript should have generous margins, preferably no less than one inch at the top and right side and no less than an inch and a fourth at the left and bottom. Double space all copy, including tables. Consider the title page as "1" and number each page successively at the top right-hand corner. Each of the following units should begin a new page.

- Title page–Include title, all authors' names, affiliations, and a line showing where reprint requests should be sent and to whom. If requested, include a running head, that is, a shortened form of the main title. (Remember to double space.)

- Abstract
- Introduction–After the Introduction, the succeeding pages of the main text may be "run on" without having each new section begin on a page.
- Acknowledgments–These usually occur after the text.
- References–Be sure to double space these also.
- Legends for figures–Double space. Use as many pages as necessary to include all the information with adequate spacing.
- Tables–Use a schematic if necessary to show how the pieces fit together.
- Appendices–Journal articles rarely include Appendices.

After the copy has been typed, read it wth more than a causal eye.* First, arrange all materials in proper sequence. Then, carefully read each page for correct spelling, correct punctuation, and agreement of reference, figure, and table citation. Do not pass by "Figure 2 shows . . ." without actually looking at Figure 2 to see if in fact it does show what you purport it to show. If you use special materials, such as formulas or verbatim translations, check each jot and tittle.

Publishers will accept minor corrections shown on copy. Merely cross out the error in the copy and neatly write the correction above the deletion. If, however, you have to correct more than two or three words on a page, you probably should correct the errors by (1) having the whole page retyped or (2) having the errors "whited out" and the corrections inserted. Correcting copy is itself a skill, and typists should learn to use the correction fluid judiciously and effectively. If you find the copy is gooky with whiting, you should have the whole manuscript retyped, perhaps by a different typist.

After you have proofread the typescript, made corrections, and proofed the corrections, have someone else proofread the copy. Because you are so familiar (if not intimate) with the material in the manuscript, you are the least likely person to notice even obvious errors. Someone who has never seen the manuscript in any of its drafty stages is most likely to succeed in spotting errors–without even trying.

*Note that we deliberately misspelled "with" (wth) and "casual" (causal).

Camera-Ready Copy

The vectors of increasing printing costs, increasing desire for editorial control, decreasing demand for high-cost printing, and the concern that advertising is more trouble than it is worth have led a few publishers to adopt printing that uses camera-ready copy. Journals that publish proceedings of meetings and a few whose emphasis is on getting papers into print rapidly use camera-ready copy—for example, *Steroids: An International Journal* and *Life Sciences.*

If a publisher wants camera-ready copy, you will be sent instructions about the exact form to be used. Usually, explicit instructions are given for preparing the first page and for setting margins. Titles, by-lines, and figures might be prepared separately by the publisher and then inserted into the copy you have prepared. Despite the established guidelines, variations occur in the appearance of articles; apparently, the saving outweighs the concern for uniformity of appearance.

Magnetic-card and magnetic-tape typing machines that store copy facilitate the typist's tasks of correcting typos, inserting changes, and adding text as the manuscript evolves through the draft stages. Although such machines facilitate the process of typing a final or a camera-ready copy, they are not required for an acceptable product. Most office typewriters can produce clear, clean copy if the letters are clean and the ribbon fresh. Electric typewriters with carbon ribbons produce an especially sharp and even type.

Accuracy of copy is possible on all machines and is a reflection of your data and of yourself. Corrections may be made with white correction liquid. Beware of correction tapes, however, which tend to flake off the copy eventually and reveal the original error.

On some typewriters, the writer can control the actual appearance of the type before it goes to the printer by using various typefaces. The copy can show italics (*behold*), Greek letters (α), mathematical symbols (Σ), brackets ([a + b]), and other elements.

ALAS . . . YOU SUBMIT

The Covering Letter

All is prepared, and the clean, well-typed copy now vibrates before you atop your desk. You are ready to send it off, to fare by its merits.

The "Information for Authors" page or the masthead tells you where to send it, how many copies to send, and most other pertinencies. Again, heed what the publisher says; the greater the adherence to the directions, the greater the likelihood of acceptance.

Accompany the manuscript with a letter, known as a covering letter or a transmittal letter, in which you say that you are submitting your manuscript for consideration of publication in that journal.

Send copies of the transmittal letter to each of your coauthors and, of course, retain one in the file with the bulk materials used to prepare the manuscript and a copy of the final draft of the manuscript.

Writing a Covering Letter: In writing your letter, consider the following guidelines.

- Date it (surprisingly, often found to be a crucial bit of information).
- Address it to the person indicated in the "Instructions."
- Include manuscript title and by-line in a "RE:" line (as in model letter).
- Explain in your first paragraph–
  (1) why you are submitting the manuscript to that journal, and
  (2) the essence of the manuscript's message.
- In subsequent paragraphs, mention–
  (1) inclusions or omissions of permissions to use photographs of patients,
  (2) financial or institutional support given to your study, and
  (3) any deviation in your manuscript from the journal's "Instructions for Contributors."
- Specify the name, address, and phone number(s) of the author with whom the publisher should correspond about all questions pertaining to the manuscript.
- Do not be overly solicitous.

MODEL COVERING LETTER

June 29, 1889

Ignatius Proteus Schmouzer, M.D., Ph.D.
Editor
Whirled-Wide Journal of Published Plethora
1542 Fourlane Avenue
Publisher Parish, Louisiana 70001

RE: "Eye-Droop, Head-Nod Confound: Correlation With Contingent Negative Variation" by Yuri Porter, M.D., Peter Doubt, M.T., Henri Crittur, Sc.D., and Gilda Lilly, Ph.M.

Doctor Schmouzer:

The results and observations from our recent study on the effects of caffeine deprivation among persons who customarily drink more than eight cups of coffee each day compare with those reported by Stuff and Such (*WWJPP* **180**:415–416, March 4, 1881), except in one important finding: whereas Stuff and Such found postprandial eye-droop without head-nod, we found eye-droop invariably coincident with head-nod.

Enclosed are an original typescript and two copies of our report of the study, which was done at Alyce Shoe School of Medicine. The manuscript is sent for consideration as a clinical note in your journal. It has not been submitted simultaneously elsewhere.

Copies of letters of permission for reproduction of Figures 1 and 3, showing patients 4 and 19, are also enclosed.

Please refer all correspondence to me at my home address, as follows:

Henri Crittur, Sc.D.
1234 Fifth Street
Bayou, LA 70304

I may be reached by phone at (office) 504-888-8878 or (home) 206-734-5432.

Sincerely,

Henri Crittur, Sc.D.
Professor of Benepractice

cc: Y. Porter
P. Doubt
G. Lilly

SIMPLER COVERING LETTER
(if you prefer)

Mary Ellen Vanderkellan, Sc.D.
Editor
Sitz-Bath Weekly
4 Tunate Turn
Wheeling, West Virginia 08080

Doctor Vanderkellan:

An original typescript and two copies of our report on the Eye-Droop, Head-Nod Confound are attached for your consideration as an article for the *Sitz-Bath Weekly.*

Please refer all correspondence to me. I may be reached by phone at (office) 304-888-8878 or (home) 206-734-5432.

Sincerely,

Henri Critter, Sc.D.
Professor of Benepractice

cc: Y. Porter
P. Doubt
G. Lilly

THE REVIEW PROCESS

The manuscript now vibrates out of your hands, your mind races off into many trajectories, toward great expectations, hope, fear of rejection, an unkindly suspicion of editors, anxiety of having to do it all over, or fantasy of it becoming the catapult to your fame ("I take as my thesis for this Nobel Oration, 'Humility . . .' "). Your time is at hand to send your manuscript off to the exacting test of review by the ritual that science has set for itself in such matters: the refereeing of articles.

The method by which experts review articles before they are officially accepted for publication is largely of North American origin. Whereas European journals have relied on the heads of university departments to critically review and referee materials that originate in their departments, American and Canadian journals have relied on persons selected by their editorial boards to decide on the value of submitted manuscripts. Receipt of an article by American and Canadian journals does not mean automatic acceptance merely because it bears a don's imprimatur. The American system probably developed out of pragmatic and philosophic values—the receipt of more articles than can be published, the rapid expansion and specialization of information in science, and the need to be fair and objective in the interests of science and truth.

An American journal usually has an editorial board of a few to a dozen or more members. For some journals the board members themselves review the manuscripts, voting "aye" or "nay" or "well, maybe, if the reserve pile is getting low." For other journals, particularly those with large circulations, the boards maintain extensive files of experts to whom they refer articles, matching an article's content with the referee's area of expertise. Ordinarily each article is sent to at least two reviewers, and their comments are heeded by the chief editor or a member of the editorial board, who decides whether to accept or reject the article. Occasionally, the member of the editorial board will reject an article without putting it into the review system if the article is poorly prepared or obviously lacks substance or pertinence to the journal. Occasionally, too, a controversial or "borderline" article may be sent to more than two reviewers.

Most reviewers take their responsibilities seriously, are competent in prescribed areas of expertise, and admit freely to inability to criticize outside those areas. Important to note is that elevation to an editorial board or editorial position is customarily through service as a referee.

That the referee system can be confounded, however, is proved by an example. A few years ago, a reputable journal published a merely theoretical, grossly questionable, hardly substantiated, singularly opinionated article that a physician had submitted "for review and consideration." Put into the reviewer's hand, it was rejected outright as "interesting in theory but not publishable" (or words to that effect). Nonetheless, the article got placed into the wrong interoffice box, became edited, typeset, and published.

In submitting your article, then, you are submitting yourself to the review process. Most journals will notify you by postcard of receipt of your manuscript within a week of receiving your manuscript. The review process, however, may take from one to six months, in which case you can only show patience in the wait.

However, if the waiting time goes beyond three months and your patience is becoming tried, you might write a courteous letter to the editor inquiring about the status of your manuscript.

The Review Decision

Once a decision is made by the reviewers and editor, you are notified of "acceptance" or "rejection" or "conditional acceptance." If your manuscript is accepted, you will receive only a letter. If your manuscript is rejected, you will receive a letter and the copies you submitted.

Skin thickness is an important measure in the writer. The need for thick skin is inversely proportional to one's success and should be distinguished from the thick skull. If your manuscript is labeled "definitely bad," an easy test of your skin thickness is whether you can accept that comment without spitting out an epithet for the one who made it. By the way, if you ask colleagues to review your draft and to tell you the truth, do not get angry or hurt when they do. Learn to appreciate their criticisms. If you can look at your own work and admit that your article is "not all that good," then you probably already have a thick enough skin.

Should that dreaded, hurtful entity—the rejection letter—arrive, read it carefully. Often comments will be included from the reviewers explaining why the manuscript did not meet the needs of that journal. Often, too, suggestions will be given for improving the manuscript–adding information, checking other references, collecting more data, or reviewing noncontributory information. If the rejection is unconditional, you should read the letter calmly, judge the value of the reviewers' comments, reevaluate the manuscript, and decide whether your original idea is still a good one. If the manuscript has value, then redesign it or improve it in whatever way necessary.

Do not grieve long over a rejection. Look at your list of other journals, and noting all the points in the "Instructions to Authors," tidy up the manuscript and send it off. Some authors do not consider an article really rejected until it has gone the round of half a dozen journals.

Sometimes the rejection letter is less than absolute: "Had this article included a complete review of the Biblical origins of xenoantibodies, we would not have rejected it." In such a case, you should consider the possibility of filling the stated deficiencies and, with a fully explanatory covering letter, resubmit the article to the same journal. However, be sure there is valid reason for hope, that you are not merely reading something into a polite twist of phrase.

Often enough, an article is accepted with the condition that certain corrections or changes be made in it. Usually these are direct and simple in the saying and difficult in the doing: "Please eliminate seven of your eight photographs and reduce the text by 50 percent." The editors do not want you to reduce the message by half, merely the text. Which one photo will you select? As with all of science, the best course of response is trial and error; see what can be eliminated without also deleting the message. If the reviewer thought it could be done, by George, you are as good a man as he is, Gunga Din. Persistence is the byword.

In all that is yours–the manuscript to live with, the editor's unabashed comments about it, the need for rebound and a substantive, resistant dermis–you are still the author seeking to become authority.

Acceptance, or Proof(s)

Hooray! The reviewers like your manuscript and the editor agrees to publish it. Congratulations, you now have more work to do–not much, but certainly painstaking work. The journal's editors will edit your manuscript, sometimes only minimally, sometimes (even after all you thought you had learned by reading this book–did you take all your dosages?) maximally.

In one form or another, you will receive a copy of the edited manuscript for review. Actual proofs to typeset material are costly; hence, many publishers do not send authors proofs. Confusingly, the edited typescript is also called "proofs," although it does not prove what the material will look like in printed form. For example, italics will appear as underlined wording on the copy. Variations in type size may be marked in the left margin by the copy editor, but these might not be obvious or understandable to the author. Delighted or dismayed, the author may find surprises such as dropped initials, unreadable blocks of "modern" typefaces, or a misinterpretation of a Greek letter (whereby a σ becomes a mere "o") in the published article.

If copy is returned to you as "proof" to be read, follow whatever instructions are included. Generally you should show any changes directly on the copy received, preferably in red. Explain difficult or complicated changes in a letter accompanying the manuscript after proofreading. *You* are responsible for the contents of the manuscript, including the changes made by the copy editor, so if you object to what has been done, you should state your objections in your return letter with the degree of strength you believe the objection warrants.

If you receive actual proofs–galley or page proofs of the printed (typeset) form of the article–follow the accompanying instructions and carefully mark any line needing a change. Stylebooks include sections on proofreader's marks, as do many dictionaries.

If possible, let several colleagues read your proofs because proofs are the final stage for eliminating error in your article. Production managers are adverse to "stopping the presses" to correct your later-found "horrible oversight," and the downtime of presses is costly. The more eyes that check your proofs, the more likely you will find every fault. Unfortunately, most journals ask that the proofs be returned in a short time, often within twenty-four hours, and having friends and colleagues read and return them "before noon tomorrow" may crimp your good intentions.

Often editors rely on return of proofs to make up imminent issues. If you delay in returning your proof, you may jeopardize production of an issue or cause your article to be delayed in its publication. Prompt return of the proofs is in your own best interest.

Delete

Reverse

Close up

Insert space

Close up and insert space

Paragraph

Move to left

Move to right

Lower

Raise

Insert marginal addition

Broken letter—used in margin

Push down space

Straighten line

Align type

Insert comma

Insert apostrophe

Insert quotation mark

Insert hyphen

Insert semicolon

Insert interrogation point

Use ligature

Spell out

Transpose

Wrong font

Set in **boldface** type

Set in roman type

Set in *italic* type

Set in CAPITALS

Set in SMALL CAPITALS

Set in lower case

Lower-case letter

Let it stand; restore words crossed out

Run in same paragraph

Fig. 19-1. Proofreader's marks.

PART VI. SOME TOOLS AND RULES: WRITER'S RACK AND 'PINION GEAR

20. Special Needs of the Writer

Beyond the pad of paper, writing implement, and time for the effort, the writer comes to appreciate other needs.

LIBRARIES

Valhalla for the writer should be his favorite library, a repository of the recorded ideas and communications of others. Increasingly, the access to those writings is facilitated by advances in the computerized technology of indexing, abstracting, and retrieval. All medical libraries belong to regional and national networks, whereby any publication available through any of the system libraries is available to the other libraries also.

On first encounter with a particular library, you should ask for a tour. Note where the specific sections are—bound volumes of journals, reserve books, audiovisual equipment, Medline services, indexes, special holdings, and special resources you may not have encountered before. You will be told what charges there are for specific services, such as use of the Medline terminal and copying machines. After the tour, look into some of the indexes to note whether you will want to make use of them. In reporting on a study involving specific chemicals used in testing biological effects, you may want to refer to *Chemical Abstracts.* If you think that an explanation for the artifact in your results may lie in a behavioral phenomenon, look into *Psychological Abstracts.* Generally, however, persons pursuing topics in medicine will refer to the *Index Medicus* and those pursuing topics in the basic sciences will refer to general indexes in their own fields.

Besides published articles, books, and current journals, however, the library has other wonderful treasures: concordances; books of quotations; references on awards, grants, and famous persons; and the special training and knowledge of the librarians themselves. With some direction from the resident experts and some initiative from you, great discoveries are possible. And they are awaiting. . . .

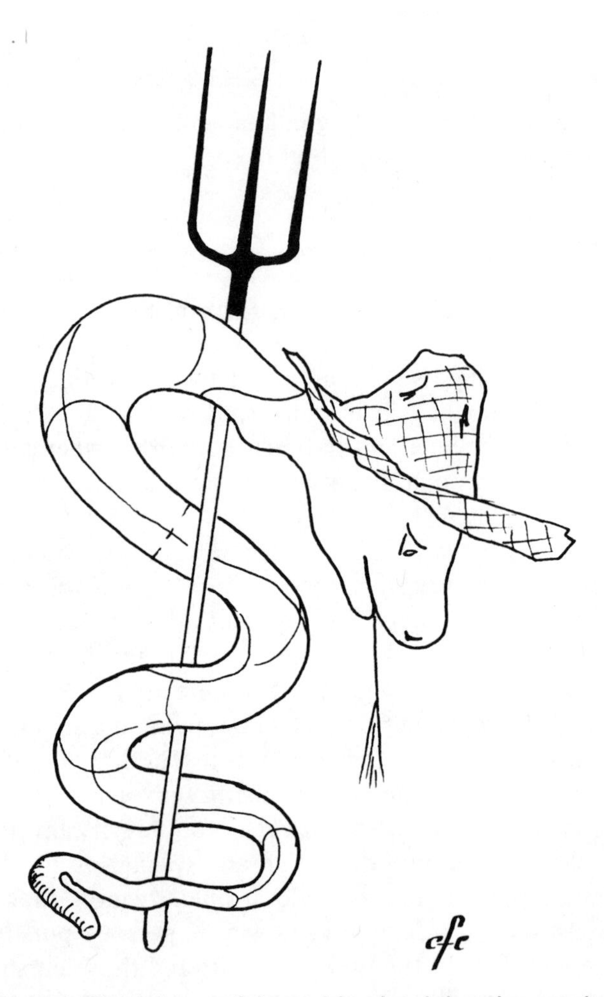

Fig. 20-1. Pitchfork: one of the special tools to help make your pitch.

TOOLS AT HAND

In the blind where you will do your quiet thinking and writing, you should have some resources right at hand for use as your pen waddles through the swamp of unrefined ideas.

Books

Dictionaries

An unabridged English-language dictionary:
e.g., *Webster's Third New International Dictionary.* Springfield, Massachusetts: G. & C. Merriam, 1971.
A good dictionary in your general field:
e.g., *Dorland's Illustrated Medical Dictionary*, 25th edition. Philadelphia: W. B. Saunders, 1974. (for medicine)
Language dictionary (if you read much in other languages)
Etymological dictionary:
e.g., Partridge, Eric, *Origins: A Short Etymological Dictionary of Modern English.* New York: The Macmillan Company, 1962.
Crowley, Ellen T., and Thomas, Robert C. (Editors), *Acronyms and Initialisms Dictionary*, 5th edition. Detroit: Gale Research Company, 1976.

Special References

AMA Drug Evaluations, 3rd edition. Chicago: American Medical Association, 1977.
Snedecor, George W., and Cochran, William G, *Statistical Methods*, 6th edition. Ames, Iowa: Iowa State University Press, 1967.
The Merck Index: An Encyclopedia of Chemicals and Drugs, 9th edition. Margha Windholz, Editor. Rahway, N.J.: Merck & Co., Inc., 1976.
Current Medical Information and Terminology. Chicago: American Medical Association, 1972.

Stylebooks

AMA Stylebook/Editorial Manual, 6th edition. Littleton, Massachusetts: Publishing Sciences Group, Inc., 1976.

CBE Style Manual, 3rd edition. Washington, D.C.: American Institute of Biological Sciences, 1972.

Manual of Style, 12th edition, revised. Chicago: University of Chicago Press, 1969.

Strunk, W., and White, E. G., *The Elements of Style.* 2nd edition. New York: Macmillan, 1974.

Style Manual: For Guidance in the Preparation of Papers for Journals Published by the American Institute of Physics and its Member Societies, revised edition, 1973. New York: American Institute of Physics, 1973.

U.S. Government Printing Office Style Manual. Washington, D.C.: U.S. Government Printing Office, 1973.

Helpful to Have Books

Pocket Pal: A Graphic Arts Production Handbook. New York: International Paper Company, 1970.

DeBakey, Lois, *The Scientific Journal: Editorial Policies and Practices: Guidelines for Editors, Reviewers, and Authors.* St. Louis: C. V. Mosby Company, 1976.

A Thesaurus

e.g., Berrey, Lester V. (revising editor), *Roget's International Thesaurus,* 3rd edition. New York: Thomas V. Crowell Company, 1962.

Books on Language Use and Writing

Cain, Thomas H., *Common Sense About Writing.* Englewood Cliffs, N.J.: Prentice Hall, Inc., 1967.

Crews, Frederick, *The Random House Handbook*, 2nd edition. New York: Random House, Inc., 1977.

Hodges, John C., and Whitten, Mary E., *Harbrace College Handbook*, 8th edition. New York: Harcourt Brace Jovanovich, 1977.

Hoy, James F., and Somer, John, Editors, *The Language Experience.* New York: Dell Publishing Co., 1974.

Irmscher, William F., *The Holt Guide to English: A Contemporary Handbook of Rhetoric, Language, and Literature.* New York: Holt, Rinehart and Winston, Inc., 1976.

McCrimmon, James M., *Writing With a Purpose*, 6th edition. Boston: Houghton Mifflin Co., 1977.

Shanker, Sidney, *Semantics: The Magic of Words.* Chicago: Ginn and Company, 1965.

Stone, Wilfred, and Bell, J. G., *Prose Style: A Handbook for Writers*, 3rd edition. New York: McGraw-Hill, 1977.

Articles/Journals

Archer, John D., Attributes of a rejected manuscript. *JAMA* **232**:165 (1975).

Beatty, W. K., Searching the literature comes before writing the literature. *Ann. Intern. Med.* **79**:917 (1973).

Blanchard, Duncan C., References and unreferences (letter). *Science* **185**:1003 (Sept 20, 1974).

Crichton, Michael, Medical obfuscation: Structure and function. *New England J. Med.* **293**:1257–1259 (Dec. 11, 1975).

Garfield, Eugene, Citation analysis as a tool in journal evaluation. *Science* **178**:471–479 (Nov 3, 1972).

Gutterman, Jordan U.; Rossen, Roger D.; Butler, William T.; McCredie, Kenneth B.; Bodey, Gerald P., Sr.; Freireich, Emil J.; and Hersh, Evan M., Immunoglobulin on tumor cells and tumor-induced lymphocyte blastogenesis in human acute leukemia. *New England J. Med.* **288**:169–173 (Jan. 25, 1973); revision by Barbara J. Culliton, **288**:173–175 (Jan. 25, 1973).

Hussey, Hugh H., Medical jargon. *JAMA* (*Editorial*) **235**:1149 (March 15, 1976).

Ingelfinger, F. J., Twin bill on tumor immunity (editorial). *New England J. Med.* **288**:211 (Jan. 25, 1977).

The editors of *Science:* Instructions for Contributors. [A mini-writing course in itself!], last published in vol. 189 (July-Sept.) 1975, after the volume index (pp xi, xii).

Shephard, David A. E., Learning to write well. *Can. Med. Assoc. J.* **116**:1106–1108 (May 21, 1977).

The Journal of Irreproducible Results, Official Organ of the Society for Basic Irreproducible Research Chicago Heights, Illinois: George H. Scherr, Publisher.

Newsletters/Bulletins/Pamphlets

Medical Journalism: The Newsletter for Editors of State and County Medical Journals, Craig D. Burrell, MD, Editor. Published by Sandoz Pharmaceuticals, East Hanover, N.J. 07936.

National Center for Health Statistics, *Vital and Health Statistics Publications Series* (e.g., "Series 13: Data from Hospital Discharge Survey") Address: National Center for Health Statistics, Attention, HRA, 5600 Fishers Lane, Rockville, Maryland 20852.

The Royal Bank of Canada Monthly Letter (recent titles include "On Criticism," "Dialogue, Persuasion and Common Sense," "Honest Communication," "Science and People," "On Saying What you Mean") Montreal: The Royal Bank of Canada.

Searching Services

Many services are available today for retrieval of information in areas throughout the spectrum of science. For information on costs and services, write directly to the organizations.

Biosciences Information Services, 2100 Arch Street, Philadelphia, Pennsylvania 19103.

Chemical Abstracts Service, Ohio State University, Columbus, Ohio 43210.

Excerpta Medica System, PO Box 1126, Jan Van Galenstraat 335, Amsterdam, The Netherlands.

Institute for Scientific Information, 325 Chestnut Street, Philadelphia, Pennsylvania 19106.

Psychological Abstracts Information Services, American Psychological Association, 1200 17th Street, NW, Washington, D.C. 20036.

Toxicology Information Response Center, Oak Ridge National Laboratory, PO Box Y, Oak Ridge, Tennessee 37830.

On-Line Information Systems (at your nearest medical library): Medline-550,000 citations from more than 3,500 journals; updated monthly.

Back 72
Back 69 } Backfiles on *Index Medicus* literature
Back 66

Cancerline–45,000 abstracts from *Cancer Chemotherapy* Abstracts and Carcinogenesis Abstracts.

AV-Line: Audiovisuals on-line. 1200 references to AV instructional materials in the health sciences.

CancerProj: 10,000 summaries of ongoing cancer research projects processed by the Current Cancer Research Projects Analysis Center, at the Smithsonian Science Information Exchange.

Bibliography on Health Indexes

For information on Health Indexes, including a bibliography, write:

Clearinghouse on Health Indexes
Division of Analysis
National Center for Health Statistics
Department of Health, Education, and Welfare
Rockville, Maryland 20852.

THE PERMISSIONS LETTER

The more writing you do the greater will be your need for a form of "permissions letter," to be sent to publishers and copyright holders to solicit permission for the reproduction of articles, figures, tables, or parts thereof from works published under their copyright ownership. Of the many possible forms for such a letter, here is one.

(Your Letterhead)

(Date)

To: (Publisher's Name)
Re: (Title or article/chapter, and journal/book in which figure, table, or selection was published)
By: (Author's name from by-line, as published)

In an article titled "(Give the Title of your article)," by (list your coauthors, if any) (and) me, I (we) would like to use the following material:

(Specify the figures, tables, or selection, including page numbers, table and figure numbers.)

Suitable credit will be given for reproduction of this material.

If you have a preferred form, please attach a copy.

A release form is given below for your convenience. Please sign and return one copy.

Thank you.

Sincerely,

(Your Signature)
Your Name (typed)
Your title

* *

I (We) grant permission as requested above, provided that complete credit is given to the source.

Date ______ Signature ________________________________

Rights and Permissions

21. Glossary

That the English language is not stable may be lamented, for precision in language changes constantly, affected by twists of meaning, turnabouts of phrases, and popularizations of erupted terms. Awareness brings one to realize that "gay" today means more than merely jolly. "Porno," now reduced to "porn," has supplanted most other similar terms—where lie "lascivious" and "bawdy"? Where once "wad" held the mountaintop of meaning for "lotsa money," today "bread" has brought wad down to the loss side of the ledger. Outside of school and often within, the pupil "doesn't dig" the content, where a generation ago he would know that it was "all Greek" to him.

Every writer has peeves, whether "pet" or mere "peevies." Authoritarianism obligates a writer to have a set of peeves against inelegant expressions, infelicitous terms, and downright errors. Akin to the pressure to publish, the pressure for one to acquire peeves is inescapable. A peeve seems a negative thing, at best, but, when applied to the knowledge-search in the literature, it can be a spark of amusement in personal awareness and, hence, a positive event. Begin now to align your peeves and preferences. Here are some of ours.

above *Above* can mean *preceding* as well as in a higher place or position. "The above statement" is proper if the statement was a preceding one, although its physical position may be below what follows (as at the top of another column).

affect/effect Both words can be nouns or transitive verbs. In the noun-sense, *affect* is limited to psychological uses, meaning (1) that which tends to arouse emotion rather than to stimulate thought or feeling and (2) the fundamental controlling element in an emotional state. *Effect* as a noun means anything brought about by a cause or agent, a result; more than likely *effect* is the noun you will need. As a verb, *affect* means to act on or have an effect on; *effect* means to bring about or accomplish, as "to effect a change."

again, back Combinations of words using the prefix *re-* with "again" or "back" are usually redundancies. "Resume again" is redundant, as is "recall it back."

aim Legitimately used to mean object or purpose. "The aim of his research was . . ." In a grant application, you may be required to state your goal, purpose, objectives, and aims. The context of the directions should tell you how the various meanings are refined.

allow/enable/let/permit *Enable* means to make able. (The money enabled her to continue her education.) *Allow, let, permit* refer to things happening. *Allow* implies forbearance; *let,* indifference; and *permit,* authority with the option to prevent.

ambient Meaning "surrounding, on all sides" *ambient* has sufficient pseudo-erudite clout to seem appropriate for use as anything from an audible pause to a redundancy for existence. The word can also mean circulating, wherein ambient air would be circulating (or surrounding) air, but not necessarily healthful air. An *ambiance* is an environment or milieu.

ambiguous, equivocal If a statement is *ambiguous*, it can be understood in more ways than one. *Equivocal* has that meaning and also one of being purposefully vague or misleading, or of being uncertain or undecided.

amenable/amendable Often confused, the two words have different meanings. A person is *amenable* if he is capable of being persuaded or is submissive. An illness may be amenable to treatment. An *amendable* program is one that can be corrected or merely altered.

among/between *Among* is more collective and somewhat vaguer than *between*. *Among* is appropriate in reference to three or more things: I put it among my souvenirs. *Between* is usually limited to showing the relationship of two items, though it can be used for more than two: The counsellor saw hope for a settlement between the two parties.

amount/number Use *number* when items are countable, *amount* for total quantities not considered individual units. The number of beer cans doubled the amount of garbage collected at Mardi Gras.

analyzation Always use *analysis.*

analyze/assay/determine In *analyzing*, you examine critically or minutely. In *assaying*, you ascertain the ingredients of something. In *determining*, you settle or decide something.

and/or The battle against use of this shortened form of "either *and* or *or*" is almost lost. Traditionalists argue that the proper sentence form would be "He gave aspirin or diuretics, or both, to all his patients."

and yet Shorten to *yet.*

angry at, angry with Be *angry at* situations but *angry with* persons.

apprehend/comprehend When you catch the meaning of something you *apprehend* it. If you understand thoroughly, you *comprehend* it.

apropos Follow this single word (in French it is *à propos*) with *of*, not *to*.

anxious/eager If you are impatiently desirous of something you are *eager*. If you have anxiety and are troubled, worried, or distressed, you are *anxious.*

as Avoid using *as* for *because;* ambiguity can result. As the solution began to boil, we reduced the heat. (Does "as" mean *while* or *because* the solution began to boil?)

as regards/regarding/with regard to Use *about* or *concerning.*

as such Sometimes *as such* is needed, but often it is not–try the sentence with and without *as such.* An example of proper use: A physician as such is not an authority on politics.

as well as The number of the verb remains the same as the subject without the *as well as* phrase. The microscope, as well as its case and ancillary equipment, was destroyed. Do not use *as well as* if *and* would suffice. The microscope, its case, and its ancillary equipment were destroyed!

aspect Meaning appearance to the eye, *aspect* is much overworked. In writing try substituting these synonyms: *phase*, *side*, *facet*, *stage.*

assess In setting or estimating the amount or value of something, you *assess* it.

autopsy *Autopsy* is a noun. An autopsy is done on a dead person, but do not write that "An autopsy was done on *all patients.*" Have it done on the "bodies."

average, mean, median In dividing the sum of quantities by the number of quantities, you get the *average.* That value is also the arithmetical *mean*, although *mean* is also sometimes used to show the intermediate value twixt two extremes (mean distance). A geometric mean is the *n*th root of the product of *n* factors. *Median* is the middle number, or the value between the two middle numbers, in a progressive series of numbers.

basis, on the basis of Try substituting *because* for the overworked *on the basis of.*

based on Change this to *from*.

bi- A confusing prefix meaning either "every two" or "twice." Call your biweekly newsletter a fortnighter (if published every two weeks) or a semiweekly (if published twice each week), but do not call it a biweekly.

biopsy Remains a noun. Therefore, a patient is not biopsied; a biopsy is done on his lesion.

blood sugar Specify blood *glucose.*

can not/cannot Use the two-word form only to emphasize the *not;* otherwise always use the one-word form.

case A *case* is an instance of disease; a case is not a patient. A physician treats the patient, gives him medication, and discharges him from the hospital. The physician studies the case, records the data for the case, and compares them with those from other cases.

causation Pseudoerudite for *cause.*

cause/reason A *cause* is what produces an effect. A *reason* is a professed motive or justification, the why. Do not use the redundancy "the reason why."

compare with/compare to/contrast Appraising or measuring one item with another is to compare the one *with* the other. I compared my data with his. To compare *to* is to assert a categorical dissimilarity. She may be compared with her sister but would be compared to the moon. *Contrast* implies a comparison to emphasize differences.

comprise/compose/constitute A catalogue comprises several sections; *comprise* means to include or cover. (*Is comprised of* is wrong.) That catalogue is made up, or *composed*, of several sections. The catalogue comprises the parts. The parts do not comprise the catalogue. A whole thing is composed of parts. *Constitute* may be used as a synonym for compose.

concept/conception/idea *Idea* is the broadest term and may be applied to anything existing in the mind. Use *concept* to express an abstract notion, especially a generalized idea of a class of particulars and *conception* to emphasize that something was conceived in the mind.

concern Apparently to avoid using *is* some writers use *concern*, producing fuzzy statements such as "The difficulty concerns where to get the money." Also, concerns is sometimes pretentiously used instead of *about*.

conscious/aware You are *aware* of events about you, but *conscious* of your own feelings or perceptions.

consensus of opinion Redundant; use *consensus* when you mean something close to unanimity.

continual/continuous If it is uninterrupted it is *continuous*; if it recurs at intervals it is *continual.*

correspond to/correspond with Correspond *with* means to exchange letters with or to be in agreement; to correspond *to* means to inform or to be similar.

criteria A plural, the singular of which is *criterion.*

crucial Of a greater degree of urgency than *important* or *critical.*

data A plural; the singular form is *datum*. Increasingly, data is being used in the singular as a collective noun, much like the word "information."

deliver *Deliver* has the increasing paradoxical reality of bearing three meanings at once, without much concern expressed by vox populi. A physician delivers both mother and child. The mother delivers the infant with the help of the physician. The child is delivered. Soon thereafter, the milkman delivers the milk. And then the hospital, physician, and milkman each delivers a bill, via the mailman. Deliver us, O Lord.

deliver/provide Although whether health care services are in fact *delivered* can be questioned, efforts to substitute *provided* or *given* have not been well received, and health care *delivery* is here to stay (at least for a while).

demise *Demise* is superelegant for *death* as used in the context, "at his demise." If you mean *death*, say so.

demonstrate Unless you "prove by evidence or many examples," use *show* instead.

diagnose The physician *diagnoses* the disease not the patient.

difference/differential Under the influence of what Fowler calls the Love of the Long Word, *differential* often imposes itself as a substitute for *difference* or *different.* Use *difference* or *different* where such is meant.

due to Change *due to* to *because of* in such expressions as "Due to the faulty equipment, he could not continue." In effect, the statement says "He (is) due to . . . or else something is owed to the equipment. *Due to* is considered colloquial in this sense and inappropriate for formal writing.

employ Use *use* if that is what is meant. Use your technique and equipment and employ your technician.

end result Used in at least three forms (*end-result*, *endresult*) the word is not likely to appear in your dictionary. Use *final result* or *end product*, which is the final result of any series of changes, processes, or chemical reactions.

etc. Avoid using *etc.* after *for example* or *including.* If possible, avoid using *etc*. altogether; instead, name the elements in the series.

etiology The controversy about whether *etiology* should be limited to mean the science of causes diminishes because the added definition of the assignment of cause has become accepted. *Cause* would be preferred by unpretentious writers in "The etiology of diabetes is unknown." If you need a long word try *pathogenesis*, which means the production or development of a disease.

evaluate You *evaluate* something in finding its value or its amount, or to judge its quality. Hence, a physician does not evaluate his patient. He evaluates the patient's condition.

explore Has a specific medical definition meaning to examine by operation, probing to make a diagnosis. Because it also means to examine carefully or investigate, you can *explore* the data.

feel In science writing, *feel* should generally be restricted to mean experiencing through the sense of touch. Where *think* or *believe* can be substituted, they should be.

Fig. 21-1. Dosage 8—original drawing.

female/male As nouns, *female* and *male* designate the gender for persons, rats, monkeys, and other species. Female persons are women and girls and male persons are men and boys. If a group includes both girls and women (those eighteen or older), referring to them as females would be appropriate. If limited to one or the other, however, the preferred usage would be women or girls, whichever is appropriate. Likewise, for males, use boys or men (for those eighteen or older) when possible.

fibrillate An intransitive verb, not a transitive. Do not write "*fibrillate* a patient," but write "His heart *fibrillated*."

following Avoid ambiguities that occur with *following*. (He was first seen following an encounter with infectious mononucleosis.) Change the word to *after*.

gavage A noun. One does not *gavage* a patient. The patient is treated by *gavage*.

gut The preponderant opinion among editors in the United States seems to be against using *gut* for intestine, despite the common use of the term by writers in England. We predict the removal of gut from the shelf of indelicate usage and that gut will become generally accepted, although *guts* (as in "he's got guts") will remain outside the pale.

history If you must add a modifier to *history*, add *medical* or *personal* or somesuch, not *past*. All history is past.

homogeneous/homogenous Things that have the same structure because of common descent are *homogenous;* a mixture composed of similar elements or parts is *homogeneous.*

impact A waning "fad word." Use it only when you mean "an impelling or compelling effect"; otherwise, use *effect* or *affect.*

imply/suggest/infer To put into the mind through an association of ideas is to *suggest.* (The smell of ether suggests a surgical scene.) To *imply* something is to indirectly express it, in a word or action. (Her action implies a refusal.) The speaker *implies* something; the listener *infers* something or draws a conclusion about it.

incidence/prevalence *Incidence* means the degree or range of occurrence or effect; *prevalence* means widely existent or general occurrence. Because the incidence of influenza was high, use of the new vaccine became prevalent.

individual Used to designate a single organism from the species and the single human being in contrast with the group. Where *person* is meant, it should be used.

inject Because *inject* means to force a substance into something, you should not write, "The patient was injected . . ." Drugs, fluids, and other substances may be injected . . . into the patient.

initiate/institute Latinate words that are overused. Use *start* or *begin.*

insure/ensure When meaning to make certain or to secure, use *ensure. Insure* is what you do in getting a policy from the Good Hands People.

interface Usually a noun, *interface* means a surface forming a common boundary between two elements or disciplines. Do not use *interface* faddishly to mean *interact* or *communicate.*

less/fewer *Less* is often misused with countable units, for which *fewer* should be used: fewer test tubes, fewer persons, fewer laboratories. For abstract or inseparable quantities, use *less:* less rain, less pain.

lesser The comparative of *little* should not be used if *least* is meant, and vice versa.

like Use *such as* to introduce a series of examples and to prevent confusion with *like* as a verb and adjective.

localize Do not use *localize* when *locate* is appropriate. *Localize* means to limit or confine to a particular place or area: Drug therapy localized the injuries to the head.

lower-level The correct adjective for low office or rank is *low-level.* They were low-level government workers.

marked/markedly In the enthusiasm to find a substitute for the overworked *very*, scientists, especially medical clinicians, took up *marked* and *markedly* and will not let go. As with *very* and *quite*, marked and markedly can usually be deleted from a statement without loss of effect. How great is a marked increase? Omit marked and markedly and state specific measures, and your work will improve (markedly).

maximum/maximal; minimum/minimal Although *maximum* and *minimum* are acceptable as adjectives, we recommend using them for nouns (rose to the maximum), and *maximal* or *minimal* for adjectives (a maximal increase).

may/might auxiliary verb forms with present or future sense. The difference between the two is in degree rather than time. *May* implies probability; *might* indicates possibility but less likelihood.

medi- A prefix that does not mean "related to medicine," but rather means (as *medio-*) "mid, middle," or sometimes "half." What then is *Medi*care?

methodology Unless you mean "the science or study or methods," simply write *method* or *methods.*

militate/mitigate Often erroneously used interchangeably. *Militate* means to have influence or effect and is usually used with "against," sometimes "for." *Mitigate* means to make to become milder or less severe. (The evidence militated against him. The drug mitigated the pain.)

modality/mode *Modality* is pseudoerudite (as in "treatment modality") for therapy. Use *therapy* or state the specific therapy used. *Mode* is the way in which a thing is done, its manner, method, or form.

morbidity *Rate* is redundant when used with *morbidity*. The morbidity (not the morbidity rate) increased.

mortality Also redundant when "rate" is added.

only Meaning is seriously affected by the position of *only* in the sentence. When in doubt about its proper place, move it successively from first word to last and note the change in implication or tone.

Only he removed three feet of intestine.
He only removed three feet of intestine.
He removed only three feet of intestine.
He removed three feet only of intestine.
He removed three feet of intestine only.

oral Because *oral* means "of, at, or near the mouth," authorities in the past objected to use of terms like oral contraceptive, oral medication. In those contexts, *oral* would mean "against conception in the mouth" and "medication for the mouth." Contemporary usage includes the meaning "of, taken, or administered by mouth," and *oral contraceptive* now means one taken *by* mouth. (Otherwise the woman who takes it might well become speechless.)

over When referring to time periods, use *during* instead of *over:* "during the past four years" or "during the period from 1968 through 1972."

parameter A word given so many uses it has become almost undefinable and meaningless. Try using *measure*, *test*, *guideline*, *limit*, *perimeter*, *extent*, *variable*, *constant*, or other terms to be more explicit in meaning. If you do in fact mean "A constant whose values determine the operation or characteristics of a system" (e.g., standard deviation), then use *parameter*.

parametric A fun word to think about in such contexts as parametric statistics (or nonparametric). *Parametric* means "situated near the uterus" in medical usage.

pathology Perhaps in pursuit of an authoritarian air, medical writers sometimes use the terms "patient's pathology," or "his psychopathology." If we limit the meaning to "science of diseases," we can talk about the patient's disease or condition without jumping into jargon.

people/persons (also see *individual*) The limiting definition for *people* derives through the Middle English, Anglo-French, and Old French from Latin *populus*, meaning nation or crowd. *Person* derives from the Latin *persona*, literally an actor's face mask, and pertaining to character. Use *people* in referring to groups having common attributes or characteristics (or diseases) and *persons* to show individuality. Thousands of people died in the earthquake. Each year, 650,000 people die of heart disease. Four persons were treated, one of whom had arrhythmia.

per *Per* in statements like "100 articles published per year" should be changed to *each*, or the "per year" changed to *annually*. 100 grams per person can be reduced to 100 grams each. Do not write "as per the directive"; write "as given in the directive."

percent/percentage Use *percent* with numbers, *percentage* without numbers: five percent; a low percentage. When using numerals, use the percent sign: 17%.

possess A somewhat strong (and long) word for *own* or *have*, *possess* suggests mastery or influence over, or managing to have sexual intercourse with. "He possessed a centrifuge" means he had or owned one.

post From the prefix *post-*, meaning later or after, the word *post* has come into use (post operation) as a preposition. Usually the meaning is better carried with standard forms (after the operation, postoperatively). But, such uses as "We measured the results post infusion" are becoming common and are acceptable.

present Some authorities question whether a patient can present signs, symptoms, or even himself. *Present*, as a verb, is transitive and therefore takes an object. If a patient presents something, that is, he offers it for viewing or display, what he presents is expressed as an object of the verb, not of a preposition. Thus, "The patient presented with headache" is ungrammatical; "The patient's symptoms were headache, nausea, and fever" meets the need.

prior (to)/previous As an adjective *prior* means preceding in time: a *prior* commitment. *Previous* connotes having *occurred* before in time or order; a previous encounter. *Prior to* means before. When possible, use *before.*

prioritize A faddish *-ize* word. Use "assign priority" or somesuch.

proven An acceptable past participle of *prove*, but *proved* is currently preferred.

purity The quality or condition of being without contamination, *purity* would seem, like pregnancy, to be an either you are or you are not condition. Although to be pure is to be free of any adulterant, apparently no reasonable alternative exists for things that are slightly adulterated. Thus, solutions may be of "various degrees of purity."

quite A qualifying term that can usually be omitted. Try writing without it. A glass that is quite full is full, a youngster who is quite fat is fat.

re A preposition that means "in the case or matter of." Used by some as a substitute for *concerning*, *re* should remain as a heading in business letters and not be used in formal writing.

require/necessitate *Require*, when used to mean "to be in need of," should be tested against *necessitate*, "to make necessary." A condition necessitates that treatment be given: the patient requires the treatment.

revealed If a person revealed a black spot on his abdomen, he *had* it, unless, of course, he was being exhibitionary. *Revealed* continues the use of Latinate words where Anglo-Saxon words are more effective.

sacrificed A euphemism for killed, *sacrificed* should be "done in." From the animal's point of view, he is killed in the experiment whatever term is used, despite the efforts by researchers to find a "softer" term. An innovator used "euthanitized" in reporting the action. When possible, state exactly how the animal was killed: The animals were anesthetized. The animals were bled to death.

science/scientific (writing) In appreciating the idiomatic subtleties of the terms, and after pondering the legalistic definitions, we have opted for *science writing*. *Sciential* was also considered.

should/would *Should* means "ought to." *Would* in the conditional use is now proper for first, second, and third person. Would you do this experiment? I would be happy to.

significant Having specific meaning in science, *significant* should be limited in science writing to the meaning of relationship to the statistical probability. Unless the meaning is obvious when such use is not intended, avoid the term and use *important*, *great*, or some other reasonable qualifier.

since/while Both *since* and *while* carry the sense of time. You should be careful that ambiguity does not result from such statements as "While the grant met his needs, he could not get help elsewhere." Usually changing since to *because* and while to *whereas* will remove the ambiguity.

some/about *Some*, often used with a numeral to mean *about* or *approximately*, does not mean *about*. "About 2,090 drugs remained on the approved list. Some barbiturates were removed from the list."

Step 2 for Dosage 7.

On the page facing the drawing (to the right presumably), write down, step by step, directions for reproducing your design. Number the steps and give only one directive in each step. Never name the design that is to be drawn; for example, do not say "square," "bird," or "streptococci." (Do not look ahead to see the next direction.)

subject The pervading influence of experimental science on clinical activities has placed *subject* on the shelf of acceptance for use as patient, person, or animal, as long as the term is in reference to an experimental situation.

sustain/maintain *Sustain* carries the stronger meaning of "having the burden of." You maintain a piece of equipment; you sustain a life.

terminate An experiment that terminates, ends. Use *end*.

there Too many writers begin too many sentences with "there is" or "there are" or "there was" or ". . . will be." Often *there* merely is an expletive, substituting for the subject, which gets pushed back in the sentence. "There is ample room for improvement" means "The room for improvement is ample."

think *Think* is used by writers who are hesitant about committing themselves to definite statements. We recommend you *believe* oftener than you *think; believe* is more assertive and assured.

this Be as specific as possible in your writing. *This* may lead to confusion if you have referred to more than one idea or thing and then refer to "this." Overcome the problem by inserting a term after *this.* This conclusion is justified. When referring to your own study, use *our* or *my.*

thrust A popular Freudian crutch. Use *objective*, *purpose*, or *emphasis*, as appropriate.

treat By definition, you can treat organisms (subject them to some process or substance) as well as patients.

unique Meaning "having no like or equal," *unique* should not be written where *unusual*, *peculiar*, or another less limiting term would suffice.

upon Can be changed to *on* in nearly every sense. (The result depends on the reliability of delivery.)

using In *The experiment was done using CaCl*, change *using* to *with*, or better, change the statement to read *CaCl was used in the experiment.* The participle *using* seemingly modifies an unidentified subject. (The experiment per se did not do the using.)

utilize/utilization Don't jump on the bandwagon of quasierudition. Use *use* or *usage.*

vary/various/varying *Vary* means "to change" (transitive) or "to show change" (intransitive). By *varying* a procedure, you change it. *Various* means "of different kinds," and *various* experiments differ among themselves, whereas *varying* experiments are ones that change or are changed within themselves.

very Paradoxically, use of *very* does not usually add understanding of the intensity or amount. A very long distance is a long distance that may seem "begged" somewhat by the *very.* Drop *very* from your writing whenever possible.

visualized The fact of something being *visualized* is usually equivalent to your seeing it. If the liver was *exposed* to your eyes, you saw it. Properly used, *visualize* means to form a mental image of something not present to sight.

whose The possessive form of both *who* and *that*, *whose* can have an object or thing as an antecedent.

wish/want You have probably used up your three wishes and still have many wants. *Want* has the finer meaning of an expression of need. Consider this string of terms: *crave*, *desire*, *wish*, *want*. Wish for success, but make your wants known.

with This innocent preposition is the source of much ambiguity. Change *He observed the patient with jaundiced eyes* to *He observed the patient's jaundiced eyes.*

22. The Can of Worms

WRITING AS BEHAVIOR, BEHAVING AS A WRITER

Whether you want them or not, success in publishing brings recognition and some "responsibilities." Before that wonder-such descends on you, know that you should be ready to defend every statement in your article, chapter, or book, and be able to accept the cutting edge of critical comment sanctioned by the world's educational systems today.

Be careful in selecting references, for many who know your subject matter will also know the reference sources. Imagine the pain of the published author who must explain misquotations, who cited as seminal an article recently proved erroneous, or who omitted the latest "hot" article that discombobulates his own data. Other distresses are overlooked errors in data (for example, showing that the *P* value was only less than 0.1, not 0.01), failure to include the key curve in the principal figure, failure to mention that half of the subjects were dropped from the study but inclusion of some of their data in the conclusions, or "forgetting" to mention the slight hedge on the data counts. As author, you become defendant. The court is the review system, but the jury is the readership. Have your brief ready, and keep a flexible profile.

Demands for Authority (On Becoming an Authority)

Your "authority" status stretches into many areas, the stretch resulting from a "pull" from various sources. Those who for penance or other inexplicable reasons assume the roles of editor or publicator of newsletters, local bulletins, or special publications, inevitably need to procure manuscripts, and the word of your having succeeded beyond the barrier of a bona fide editorial board immediately gives you entry into the less-hallowed ground of parochial professional politics. "Write an editorial" or "How about a short commentary?" or "Just dash off about ten pages" will come the plea, and assertiveness borne by the confidence of being published may well cause your head to nod.

Regional, state, and local chapters of your professional groups need speakers, speakers, speakers. Somehow the system believes that speakers are "in-the-flesh writers" who, having proved that they can put the shot into publication, must surely find it no trouble to tell the audience the why and wherefore of that put. You will, of course, be expected to do more than simply read what you have already published, so keep all your notes, doodles, lab books, printouts, and memos—you may well find content in them that would be more meaningful to an eye-to-eye audience than to a readership.

Beyond the call for local publication and some speaking assignments, appreciate that you may well be beckoned toward other titles. If you published an article on bees, then the National Beesness Association may well invite you to become a member of its Board of Directors. Remember that, having published, you become a de facto authority. You might find that some fulfillment could come with working as an official for one of the many organizations established to meet society's needs with specific problems. If you can find a group that fits your own interests, the investment you make may further serve your purposes in research and writing. Selection of subjects for testing, study management, and even study design can be facilitated through organizational planning and know-how. Ultimately, of course, you may become the director of the group, whereby your need to communicate will grow, and the obligation to publish and speak will grow boundlessly. Such ultimates should be thought of as you set your course. Opportunities occur; you only need to ask yourself if you want to respond to them.

Authority increases with each success you have in publication. In each successive manuscript you prepare, as you report, "In a previous study ([your name], 1978) . . ." your authority is proved, if not taken for granted.

The proof becomes reinforced by other means. Others writing in your subject area may soon begin to quote your article, or cite it for corroboration or rebuttal, or just to show an awareness of current activity in their field. Some even seem to play the game of "cite me and be cited." A comparatively new resource, the *Citation Index*, will, in fact, report who has cited your work.

Your published article will carry implications for your institution; the fact of your authorship will be part of its manifest image. "Smith at Dale, and Jones at Yuke have studied that . . ."—so will go the conversation. But, beyond that, your publication is justification in applying for grant money (also of concern to your institution); the request for a list of "pertinent personal bibliography" is not merely cursory for the grant-givers.

Rightly or wrongly, promotions still hang on the fact of publication. Proof of one's recognition in a field is the list of his publications; the longer the list the greater the proof. Department heads worry about the image of their departments within the image of the institution as a whole. Often one's bibliography is as significant as his academic training in the continuance of his departmental activities. And beyond the institution itself, beyond the department, is your community, which has its own need and demands for notables. Fame is a force of subtle portent.

THE NOT-SO-ANCIENT RITE OF COPYRIGHT

Who, you may well ask, should have the "exclusive, legally secured right" to reproduce and publish the very work you have written? "Why, *I!*" could be your reasoned retort. And, in many instances, the right could be yours. Authors typically, though not always, retain the copyright on books even though they contract other rights—to publish, to distribute, to film, to record.

Copyright protects the things a person creates—original forms of written works, songs, paintings, and the like—but not the ideas per se, or even their titles. Throughout the world, copyrighted materials are regarded as a form of property, assuring protection of the created work for the originator to control his copy (in written materials, the actual sequence of words) and to copy (that is, *reproduce*) his work. The right is a personal one that the author retains as a means of preventing distortion of the work, of identifying himself as the author, and of gaining profit from the work.

Until 1710 in England, when the Statute of Queen Anne was enacted by Parliament, anyone could claim the works of a contemporary as his own for whatever ends. Monarchs sooner or later saw the revenue advantages in charging authors for registering and licensing their works, and copyright laws eventually were passed in Denmark (1741), France (1793), Spain (1847), and Germany (1870). In the United States, by 1786 all states except Delaware had copyright laws. In 1790, a federal copyright law was passed. After centuries of political manipulation of publishing and expansion of printing technology, a convention was held in Berne in 1887 to set the groundwork for the International Union of all the important European countries. Member countries granted foreign authors published elsewhere the same rights given their own citizens. Foreigners were granted rights by the United States in 1891 and later, more thoroughly, in 1909 by the Copyright Act.*

In the United States, effective January 1, 1978, copyright has been according to Public Law 94-553, signed on October 19, 1976. Formerly, unpublished manuscripts were protected automatically by common law at the state level, and published manuscripts were copyrighted and registered under federal statute. Public Law 94-553 establishes a single system of statutory protection for all copyrightable works, whether published or unpublished. The new law provides copyright protection to works of authorship for the author's life plus fifty years and does not require renewal for works registered after January 1, 1979. Information can be obtained from the Copyright Office, Library on Congress, Washington, D.C. 20559.

The Journal Editors' Jump

When you submit an article to a science journal, you may elect to retain the copyright yourself or you may transfer it to the journal. Transfer of copyright ownership can only be accomplished by signing an instrument of conveyance, and that transfer by the author must be voluntary and without coercion or covert pressure. If you choose to keep the copyright of a journal article, you must deposit with the Copyright Office within three months after the date of publication two complete copies of the printed edition along with a completed Application for Registration of a Claim to Copyright and ten dollars. You will receive a Certificate of Registration with an official seal.** If you transfer the copyright to the publisher, permissions for reproduction or further use of the work must then be obtained from

*Margaret Nicholson, *A Manual of Copyright Practice for Writers, Publishers and Agents*, 2nd ed. New York: Oxford University Press, 1970, pp. 3–7.

**Don Glassman, *Writers' and Artists' Rights*, Writers Press (2000 Connecticut Ave., Washington, D.C. 20008), 1978.

the publisher. Your permission (as author) may be asked, but such involvement is only through the courtesy of the soliciting person. Even if you decide to reproduce it somewhere, you must first get the permission of the publisher because he holds the copyright.

A Slice from Another's Pie

With the persistence of publication, today's information necessarily builds on works of the past. If you decide to use some portion of another's work in your article, how much can you use without infringing on that author's copyright? Try as it will, the wisdom of man has not found anything better than the "doctrine of fair use" to guide you in such use.

> *Notwithstanding the provision of section 106, the fair use of a copyrighted work, including such use by reproduction in copies or phonorecords or by any other means specified by that section for purposes such as criticism, comment, news reporting, teaching (including multiple copies for classroom use), scholarship, or research, is not an infringement of copyright. In determining whether the use made of a work in any particular case is a fair use, the factor to be considered shall include–*
>
> *(1) the purpose and character of the use, including whether such use is of a commercial nature or is for nonprofit educational purposes;*
>
> *(2) the nature of the copyrighted work;*
>
> *(3) the amount and substantiality of the portion used in relation to the copyrighted work as a whole; and*
>
> *(4) the effect of the use upon the potential market for or value of the copyrighted work.* (P.L. 94-553, sec. 107.)

The new law recognizes the principle of fair use as a limitation on the exclusive rights of copyright owners and indicates factors to be considered in determining whether particular uses fall within this category.

The guideline, and it is only that, for use of another's words in your own work has been to limit such use to fewer than 250 words. Beyond that number, be sure to get permission for use of the quoted material. Note that, without permission, you can paraphrase or rewrite the material in your own words, as long as you give credit to the original author for the material. A footnote is a standard and simple way of acknowledging the source. If you have any doubt, however, you should write to the copyright holder and get permission for use of the published portions you intend to quote.

A journal publisher from whom you have requested permission to quote a work may ask you to also get permission from the author for use of his work. Such permission may apply to reproduction of figures and tables as well as to published written passages. Publishers and authors are usually cooperative about granting permission as long as the intended use of the material is noncommercial. Some publishers receive hundreds of letters throughout the year and ordinarily stamp the permission directly on the letter, which is then signed by the publisher's agent and returned to the person requesting the permission.

MAKING IT BY FAKING IT

Publishers, authors, and readers have a tacit set of ethics for the one who offers an article for publication. Do not cite references you have not read. Do not report as an observation something that was not observed. Do not claim another's data as your own. Do not offer your article to more than one journal at a time!

Some authors are willin' villains. Many articles have been written in the past few years about persons who have been found to hedge their data, to use instinct as a guide to their proofs (with some spontaneously generated data to support the instinct), or to extrapolate meanings from their studies far beyond reasonable conclusions. Sometimes a researcher responds to indirect pressures to provide progressive findings, and, sensing the results he is going to get, uses inexplicable means to "obtain his data."

Any author who has done much research and reading in his special area of interest has encountered the problem of having to decide what his course of action should be in a "delicate situation." What do you do on finding a small error in a critical set of published data? If the last rat ran through the maze in 15.7 seconds and would have put the P value at 0.01 if only he had done it in 15.4, "timing error" becomes an important consideration. Unfortunately, at least a few researchers might use or overlook the published data or "find" that the last rat in fact ran the maze in 15.4 seconds to prove the hypothesis. In such an event, one might conclude that the last rat was the researcher himself.

Motivation to follow a system of ethics can often be inspired by department heads, who control the flow of ideas within their disciplines and the flow of research, grant proposals, and publications from their departments. The best stopgap in the publication process, however, remains the reviewer who is alert to the activities in his specialty area and who can discern the good and original works from those that are not. Ultimately, of course, truth itself is the most pervasive stopper of deceit, although it may not be discerned in time to serve as a stopgap.

Freedom of Information Law

Priority of discovery has become one of the worms in the can that must get its wiggle's worth. Because the Freedom of Information Law has enabled many of the formerly closed avenues of communication about scientific research to become open, priority of discovery customarily established by the date of publication of the research may become a thing of the past. Still, if Smith does forty-seven years of research on the libidinous function of the uvula, and Krghnghouiey publishes an article on it after three weeks of reading and two days of yawning before a mirror, Krghnghouiey will have his name affixed with the findings. Smith will play second fiddle, and most likely the more he would argue about priority of his findings the more he would be suspected of "protesting too much."

Thieves, 'tis said, are everywhere, and *Science** recently published a report on the fact that the Freedom of Information Law is apparently allowing some institutions to get wind of what is blowing about in specific research areas and using that information to prepare their own grant proposals. In other words, Institution A, having gone through the labor of working up a definitive grant proposal and finding it rejected (for whatever reason) by the granting institution, soon learns that Institution B has submitted the same proposal but with refinements to adjust for the reasons for rejection.

*Philip M. Boffey, "Grant applications: Panel finds new laws enable stealing of ideas," *Science* **193**:301–303 (1976).

Playgermania

In those ancient times when egos were in need of attention and Ellipses filled his product-void by taking as his own the works of Parentheses, Ellipses was merely a *plagiarius*, an abductor (in this case, of words). Today, he would be shown less indifference; having gone beyond "fair use," he would find his destiny channeled through the courts if his thievery of another's work were serious enough to challenge, and he would surely suffer the finger-pointing of outraged peers if they learned that pickings were more than inadvertent duplication and were accompanied by no evidence of remorse.

Federal Fakery

Perhaps our concern today is set too much on matters that pertain to numbers of words or pages in "documents" and too little on what real substance exists within those writings. The acknowledged (indeed, self-acknowledged) leader in the field of using "wordese" to provide an ostensible message when no real message exists to be given is the government, at all levels. In *The Washington Star*, a regular feature called "Gobbledygook" includes excerpts from laws and other documents. Here is a sample feature published March 16, 1976.*

A-10 The Washington Star Tuesday, March 16, 1976

Gobbledygook

Section of a tax law, quoted in a Labor Department directive on "Deferral of Federal Unemployment Tax Credit Reduction" for taxable years 1975 though 1977:

"The provisions of the preceding sentence [i.e., prescribing tax credit reduction,] shall not be applicable with respect to the taxable year beginning January 1, 1975, or any succeeding taxable year which begins before January 1, 1978; and, for purposes of such sentence, January 1, 1978 shall be deemed to be the first January 1 occurring after January 1, 1974, and consecutive taxable years in the period commencing January 1, 1978, shall be determined as if the taxable year which begins on January 1, 1978, were the taxable year immediately succeeding the taxable year which began on January 1, 1974.

Fig. 22-1. Gobbledygook.

*Reprinted with permission of *The Washington Star.*

Dosage 63: Gobbledygook

Write your interpretation of that Gobbledygook statement (and *then* see our interpretation in the Appendix).

With legitimate research being "approved but not funded," the dismay of serious researchers is understandable each month when Senator Proxmire announces his Golden Fleece award, given to that person or agency that has been awarded money for the most ridiculous research exercise. Generally the research seems to be such in name only, and the design or purpose seems to acknowledge imaginative blatancy. Following suit, perhaps after receiving a rejection on a grant proposal, the researcher intending to study the relationship of a particular organism to heart disease should resubmit it as "Heart Shimmy Strep, Do They Hear the Beat?" . . . or somesuch.

At any rate, government seems to finally be aware of its influence on using many words to say nothing (1) in having set up the Commission on Federal Paperwork; (2) in President Carter's stated objective of having "understandable language" used in the preparation of reports, memorandums, and documents; and (3) in other governmental bodies (for example, the Pennsylvania legislature) deciding it is time the people were able to understand that which was otherwise hidden in the legalese, governmentalese, and the "ese-y" writing of all professions. The only danger, seemingly, in having laws written so that a high school graduate can understand them is that special interest groups would find it difficult to justify their causes. Were laws written with communication as their objective, perhaps interpretation of the law would be less a matter of judicial opinion and more of a juridical activity.

Ghost Writing

Hiring a writer to produce something to which another will put his name as author is fairly common. Usually such activity is excused with the comment, "I know what to say, I just don't know how to say it." Ideas are poured from the mind of one person into the mind of another, a ghost writer, who then does the writing. Using a ghost writer, for whose services you would pay, raises the ethical question of buying fame. Increasingly, such writing is becoming less ethereal.

Proper credit should be given to the person who actually does the writing. Such acknowledgement may be a simple line at the end of the manuscript ("Penwith Inkdip assisted with the writing of this manuscript.") or it may be shown by the inclusion of the writer's name in the by-line. A few publications now allow a "sub-by-line" for inclusion of information about assistance with the writing of the manuscript.

Some writers have such skill with words that they can make a manuscript give the work implications it should not carry (but which would increase the likelihood of its acceptance). The person hiring such a writer has ultimate responsibility for the content, including the defense of statements made that could not be corroborated or defended under scrutiny.

DEMONS

Demons crawl around the terrain of the science publishing world. Because of the number of journals, and great demand each week, each month, each quarter, each year for content to fill the published pages, the demons find easy tread on that terrain. The demons, of course, may be seen in a different guise, depending on which side of the publishing fence one sees them.

Publisher Whims

The mix of publishers, each having to decide questions of style, format, objectives, and audience, gives rise to other demons. The editing demon plays subtly with your manuscript, converting your abbreviations, misreading your Greek alpha as a script x, and lowercasing those very terms you took great care to capitalize. Yet, *you* are responsible for all the content, including the editor's changes.

Some demons work within the publisher's procedural structure, positioning your manuscript in the journal next to an ad or another article that proves embarrassing to you. The truncated title in the list of contents may show your article to be emphasing something you had not intended. Your figures may be printed upside down, cited in the wrong order, or have the wrong legends.

The ominous demon that has become most brow-wrinkling is the one in charge of page charges. To recoup some of the costs they

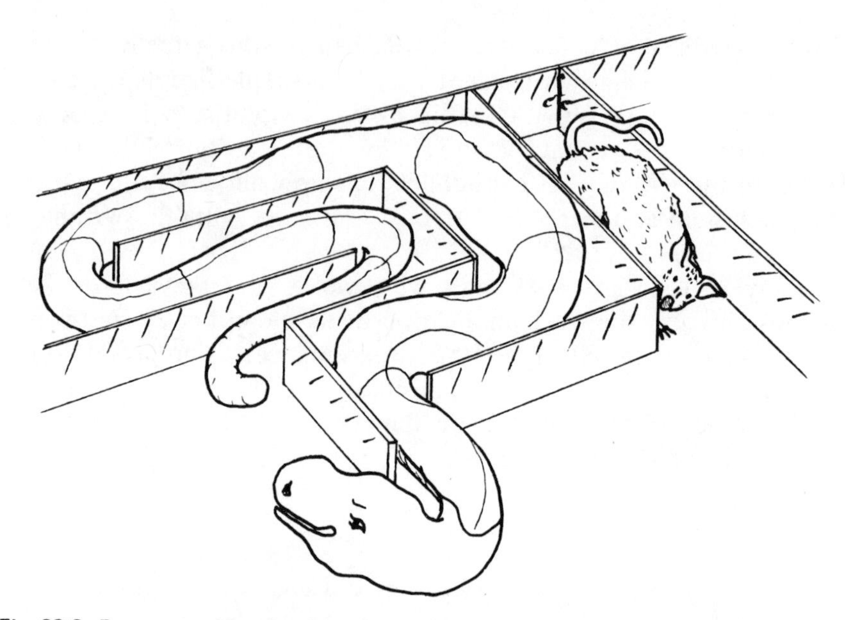

Fig. 22-2. Be sure you have the right goal in sight, are not just chasing your tail, and know when to be amazed.

bear, some publishers charge authors per page to publish their articles. Some charge a set fee for each page; some charge only for the number of pages beyond a specified minimum (for example, for each page or portion thereof over four pages). The demon in charge of page charges will be busy for a long time indeed.

Advertising

Of the thousands of journals, many must strive to survive the onrush of cost upslide, the insult of income downturn, and the ever-present ogre of rising competition. Advertising was once the great golden goose that paid the way for journals and organizations as well (formerly, for example, to pay the cost of AMA's publications and its programs, whereby no dues were needed for members). Some embarrassment has arisen in the relationship of journals and advertisers, to the point that some have considered dropping all advertising and letting the support derive from subscriptions alone. Ultimately, the cost in such situations necessitates changes in production means, frequency of publication, audience, and even content. The financial success of the so-called throwaway journals, which rely on advertising for their profit, is a contrast to the state of affairs for most noncommercial journals.

Advertising is cloaked in the ostensibility of giving information to journal audiences. For the privilege of such communication, the advertiser is willing to pay for insertion of his ads in the journal. In fact, if the journal opens its pages sufficiently for such advertising, advertisers may be willing to pay enough to cover the costs of publication–for the whole publication. Some journals realized the advantage of advertising economics before others, but undoubtedly the "holler for dollars" has led many to consider it more seriously in recent years.

The influence of advertisers on the editorial policy and contents of a given journal is conjectural, but their influence on the arrangements of those contents is certain. For many years the ads were limited to specific sections of journals. Usually they were clumped as "front" and "back" ad sections. Persuaded that "interspersing" ads would be of advantage–the advertiser would be willing to pay a premium to have his ad appear smack among the regular science articles–some journals took the leap and spread the ads throughout and amongst the pages. The disdain of the readers was substantial but not well expressed or organized, and the effect on the anatomy of the journals remains. That that effect has not yet seriously impaired science content, however, is shown by the fact that occasionally an ad will appear juxtaposed to an article written that refutes the product's use or questions its efficacy.

A writer should be aware of the kind of advertising a journal contains because it is likely to be a good guide to the kind of objectives the journal has. If your article is about drug efficacy and the journal carries no drug advertising (but does carry some advertising), you may be reasonably sure that the drug advertisers did not consider the journal worth advertising in or that, for some reason, the journal itself has a policy against such advertising. In either case, the question of drug efficacy is certain to be a sensitive concern for that particular journal.

Reprints

In the days before photostatic copying, reprints (also called offprints) were a consuming business for authors, secretaries, spouses, and librarians. "Here, send for a reprint of this article, Hon." Most authors ordered from fifty to several hundred copies of their articles, and then jammed the excess into the bottom file cabinet drawers for ballast and occasional happening on.

The facility of photocopying has changed the practice of ordering reprints from authors. Most interested readers would rather not wait for the languorous mail service to bring them a formal reprint; they prefer to send the office helper to a photocopy machine immediately.

Today, reprints are a business unto themselves, and often the original publisher does not fill your reprint order; that is done by a reprint house, which only indirectly is affiliated with the publisher. Demons romp in the domain of reprint houses. Any problems you have with reprints will be "out of the hands" of the original publisher. When your check gets lost or you find your reprints have been shipped to Guam, you will deal with the reprint house, most likely.

A common practice is to take unbound pages from an issue and distribute the "tear sheets" to the authors whose articles appear on those pages. Thus, you as an author would receive a batch of such tear sheets, which are pages *as printed directly for the journal;* hence, they might include the end of one article and the start of your article or the start of another article at the end of yours. Many authors consider the tear sheets sufficient to their needs.

Occasionally a commercial firm (pharmaceutical companies, for example) finds a published article to be "of educational value" and orders reprints to distribute with information about one of its products. Such distribution is ethical as long as it is truly "educational"; if the company decides to change the content of the article in any way, the distribution would be unethical. If any request is made to you for changing the article after it appears, consult with your publisher about the propriety of the distribution.

Sometimes (for ambition takes many forms) firms will ask that "preprints" be made of your article—that is, that it be printed before the journal actually publishes it. That would make the article available for mass distribution at the time the journal itself actually goes into the mail pouch, if not before. Again, you must consult with the journal publisher about such practices; if the article is distributed before the journal is printed and distributed, the publisher's copyright is violated.

Perhaps the most tested reprint will be yourself. Once your article is published, you will be expected to know every word and phrase in it "for aye." The responsibility extends from being an "author" to becoming an "authority," because all *published persons* are, by common unconscious agreement, authorities. Wear your mantel solemnly.

HORIZONS

Besides the reward of fame that comes with publishing articles and the kudos that come from colleagues, you may also, with a touch of luck, be eligible for an award or two. Many organizations and institutions give awards for the best article or book to members. Some are competitive and have prescribed rules and topic areas; others are less prescriptive. Information on such awards is available at libraries and the headquarters of the sponsoring agencies.

Should you try incorporating yourself if you seem to have a lot of success as a writer? The question of incorporation should be one only professional writers consider seriously. It involves many advantages and disadvantages that depend on one's state of residence, income status, need for liquidity of funds, commitment of capital, and profit and loss obligations. The most sound advice for the person intending to incorporate himself is to get sound advice from a sound advisor.

HAPPINESS: GOING FISHING WITH YOUR CAN OF WORMS

Whatever is said about writing it has this advantage: it obliges the writer to think about whatever topic he has in mind to write about. Often the writer will be amazed at the relationships that come as products of logically ordering information. The reward to the writer is that of the joy of discovery, the happiness in having found the proper words to express a point, and the ineffable feeling of contentment in seeing his words put on the page.

Playing the line of being a writer can be both hobby and business; but, whatever attitude you take, know that you can get hooked on the activity. The real business of writing gets down to communicating with yourself and with others who will find help, comfort, or pleasure in what you render on the printed pages.

Publishing is still the means to avert perishing in academia, to have some assurance for promotion, to acquire fame (but usually at the sacrifice of fortune), and to align one's destiny with the Notables of the Ages. Inasmuch as happiness lies in such pursuits, you should cast about, not let your lines go un-a-baited, and use your can of worms to full advantage.

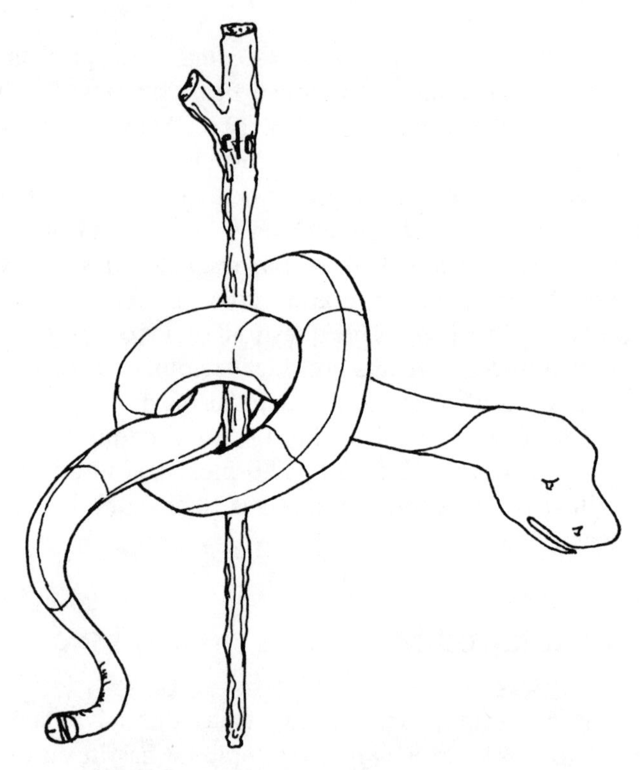

Fig. 22-3. Don't forget to write.

Appendix

Responses to Dosage 9

(1) ? The head of the school found three technicians playing cards, who may have been men, or men and women.
(2) ? The head of the school walked in; the head (perhaps the Chancellor) and the dean are not necessarily the same person. Nor do we know that he walked in; he entered.
(3) ? Policy did not forbid gambling, but we do not know about playing cards.
(4) ? We know that Dr. Henry Watson is a faculty member, but he is only one of many faculty members, so we cannot assume that it was his laboratory.
(5) T or ? Since gambling is not specifically forbidden, it probably is not punished–unless by insidious means.
(6) F
(7) ? Again, we do not know if the head and the dean are identical.
(8) F
(9) ? They were playing cards, but not necessarily gambling.
(10) ? Were they surprised? Perhaps the school's head was the fourth player.
(11) T
(12) ? We do not really know; he could have expressed himself forcibly for or against gambling.
(13) ? Was it in Dr. Watson's laboratory?
(14) ? The school is not necessarily a medical center.

Responses to Dosage 10

(1) T
(2) ? How *did* he assess Harris?
(3) F
(4) ? Was he one of the three department heads granted the privilege?
(5) ? At *least* three had the authority . . . .
(6) T
(7) ? Three *men?*
(8) ? Maybe five.
(9) T
(10) ? Were they the *best* department heads?

Responses to Dosage 18

The function examined here is . . . .
Freud contributed greatly . . . .
In addition, Freud conceived . . . .
Although the idea of death implicates metaphysics . . . .
We attempted to define and describe . . . .

utilization–utilize	use, apply
manifestation–manifest	show, evince, evoke, evidence
development–develop	have, show, grow
identification–identify	name, label, call
demonstration–demonstrate	prove, show, evidence
consideration–consider	think, believe, weigh, know
illustration–illustrate	display, show, draw
separation–separate	divide, cleave

Responses to Dosage 19

(1) Reduced sentences
When you exercise, firmly grip the lever that activates the winch.
New liver cells replace the destroyed cells.
If you cannot observe enough timed trials, randomly select other subjects.
How soon will the animal respond?
About 16 million Americans have gallstones.
She replaced the left hip with the new prosthesis.
The diagnosis was chronic gout.
The study's goal was to develop a method for modifying human malignant cells.

(2) Invigorated Paragraph
Table 1 includes data on the endoscopic complications, which varied with each technique. Today surgeons can dilate the esophagus by three methods: mercury-filled rubber bougies, metal olives passed over a guide-wire or string, and pneumatic balloon. The complication rates for instruments designed by different manufacturers for the same procedure did not differ significantly.

Responses to Dosage 20

obviously
to continue

in midafternoon, at 3:00 P.M.
at midday, at noon
incubates

In these studies, the exogenous administration of the drug did not mimic the effect on glucose.
The test apparatus may have been defective.
We believe the rough texture and the scratchiness of the sound when the material was scraped with the thumbnail led to his distress.
Most physicians believe the treatment is efficacious.

Response to Dosage 21

(2) As part (which part?) of an ongoing (forever?) interdisciplinary (which disciplines?) study, we set out to measure the effects (what kind?) of exercise (what kind?) on obese persons (adults? men? women?) attempting to reduce (what? blood pressure? weight?) by dieting (how? what kind of diet?).

Responses to Dosage 26

SGOT = Serum glutamic-oxaloacetic transaminase
DMSO = dimethyl sulfoxide
IUD = intrauterine device
DNA = deoxyribonucleic acid
MU = mouse unit
mRNA = messenger ribonucleic acid
pg = picogram
PD = prism diopter
MHz = megahertz
CFT = complement fixation test

Responses to Dosage 30

(1) *With* no increase of insulin level, an increased catecholamine level would increase cyclic AMP concentration *with* later glycogenolysis enhancement.
(2) The students received the lipid study results *by* campus mail *in* their freshman and senior years *with* a questionnaire *about* any weight change.
(3) The government appropriated 10 percent more research dollars (FY 1974–77) *for* digestive diseases.

Responses to Dosage 31

After resting fifteen minutes in the sitting position
After the patient sat and rested for fifteen minutes, a baseline arterial sample was drawn.
before making any incisions
All hemorrhoids should be carefully inspected before any incisions are made.
Using a balance beam   *wearing no shoes*
By means of a balance beam, weight was measured for ambulatory patients, who were dressed in light clothing and wore no shoes.
On regular chow and using an interrupted schedule
With the animals on regular chow and on interrupted schedule, we detected practically no sterols in the plasma of rabbits.
feeding a mixture of plant sterols
However, when the animals were fed a mixture of plant sterols at 2 percent level for ten weeks, measurable quantities of campesterol were present in the plasma of all rabbits.
using metofane for anesthesia
The pylorus was ligated with triple 0 silk suture, and metofane was used for anesthesia.
Dying prematurely
When a patient dies prematurely, his family can undergo extreme distress.

Dosage 36: Simple Sense

Ten Simple Sentences
New ego functions can be named.
Such naming obviously contributes enormously to psychoanalytic theory.
Ego psychology has arrived.
Psychoanalysis has been liberated from the shackles of a strictly motivational psychology.
Autonomous spheres and apparatuses have been discovered.
Behavior can finally be understood as the operative consequence of specific ego functions.
Our view of the ego has thus expanded.
This expansion of our view of the ego has even enabled us to solve the problem of identifying the essential elements of psychoanalytic treatment.
Such treatment is now definable as the cooperation of the analyzing function of the analyst's ego.
It also is definable as the synthesizing function of the patient's observing ego.

Revised Paragraph

Ego psychology has liberated psychoanalysis from the shackles of a strictly motivational psychology. The discovery of autonomous spheres and apparatuses has led to an understanding of behavior as the operative consequence of specific ego functions. With this expanded view of the ego, we can now identify the essential elements of psychoanalytic treatment as the cooperation of the analyzing function of the analyst's ego and the synthesizing function of the patient's ego.

Responses to page 140

We shall stay *or* we shall go.
The rain falls *and* the flowers grow.
Troubles come *but* life goes on.

Responses to Dosage 38

Paragraph 1: (a) The old woman didn't change her position until (b) he was almost into her yard; then (c) she rose with one hand fisted on her hip. The daughter, a large girl in a short blue organdy dress, saw him all at once and jumped up and began to stamp and point and make excited speechless sounds.

First Sentence

a = main clause
b = subordinate clause
c = main clause

a + b: complex
b + c: compound
compound-complex

Second Sentence

Simple sentence with compound verb.

Paragraph 2: I proceeded to forget Maurice, but not this DNA photograph. A potential key to the secret of life was impossible to push out of my mind. The fact that I was unable to interpret it did not bother me. It was certainly better to imagine myself becoming famous than maturing into a stifled academic who had never risked a thought. I was also encouraged by the very exciting rumor that Linus Pauling had partly solved the structure of proteins.

First and Second Sentences: Simple sentences.

Third Sentence: Complex sentence with "that," a relative pronoun, functioning as a connective.

Fourth Sentence: Complex sentence with "who," a relative pronoun, functioning as a connective.

Fifth Sentence: Complex sentence with "that" as a connective.

Responses to Dosage 40

(1) The American Heart Association estimates that . . . .
(2) We diagnosed (the condition as) tophaceous gout.
The resident diagnosed tophaceous gout.
(3) Table 1 summarizes the data for the 312 patients.
(4) The examining physician found an enlarged . . . .
(5) Lutz was first, in 1953, to describe elastosis perforans serpiginosa. Lutz described elastosis perforans serpiginosa first in 1953.
(6) Of the different dietary fats, peanut oil induced lesions . . . .

Responses to Dosage 41

He added sal ammoniac and salts of calcium and sodium.

Activity was detected in regions *b* and *c* not only when antisera to HSV was reacted with infected cells, but also when normal sera were substituted and, to a limited extent, when reactions involved uninfected cells.

Some techniques that can be used to teach the student the art of medicine are to have him observe (1) the way the preceptor handles problems, (2) the decision-making process for hospitalizing patients, and (3) the use of community resources.

He was graduated at the top of his class, was president of the student body, was editor of the school's publication (*The Zinger*), and will be awarded the Fernal Foundation Award in September.

Responses to Dosage 42

Dependency seems to relate to smoking in much the same way it relates to alcoholism and drug abuse. Humans develop behavioral rituals to cope with conscious needs and threats. This dependency confounds laboratory data that are used to establish whether smoking is an effect of biological dependency, for example, contingent on the amount of nicotine in the blood. The evidence for dependency, whether biologic or psychological, is clearest in the fact that, although many believe smoking to be harmful and even cancer-causing, 60 million Americans smoke cigarettes.

Responses to Dosage 43

Although many believe smoking is harmful and even causes cancer, proving that it is carcinogenic for humans is difficult. Sixty million Americans have apparently not been sufficiently persuaded by the evidence to forego their habit of smoking. Whether the habit itself is dependency-related (for example, contingent on the amount of nicotine in the blood) remains an open question. Humans, unlike laboratory rats, depend on behavioral rituals and confound any conclusions that might otherwise be drawn from the data. Dependency seems to relate to smoking in much the same way it does to alcoholism and drug abuse.

Responses to Dosage 44

Paragraph 1: Topic sentence is first sentence. Paragraph is descriptive. Summary: The ravaged artwork is of unknown origin.

Paragraph 2: Topic sentence is first sentence. Paragraph is inductive. Summary: Competitive pride can increase motivation to tolerate pain.

Responses to Dosage 45

(1) Begin new paragraphs:
It requires, first of all . . . .
It was Pablo Neruda . . . .
But individual commitment . . . .
Certainly, I think . . . .
We in the schools . . . .
Whatever our ultimate strategies . . . .

(2) Begin new paragraphs:
Risks of death from industrial injury . . . .
Earnings fall with increasing age . . . .
In one South Wales mining village . . . .
Turning to specific diseases . . . .
In most disabilities . . . .
Complicated pneumoconiosis . . . .

Responses to Dosage 46

- Most likely candidate for topic sentence is the last one, beginning "The compactness and legibility. . . ."

- The topic is the accommodation of new words and new senses with retention of available words deemed includable by use of Times Roman and a larger page size.
- The third sentence ("By itself, the number of entries. . .") seems to be an aside.
- The message of the second sentence ("It would have been easy. . .") seems to be restated in the fifth sentence ("To make all the changes. . .").

Responses to Dosage 47

Paragraphs 2 and 3 give the parts of the whole, in this instance a categorization and description of the recording results. The sequence of sentences could have been based on an orderly progression along the anatomy; but, more to the point, it subordinates anatomy to the types of responses. Note that although both paragraphs describe types of recording response, those made from the cord surface are described in a paragraph separate from those made on the skin level.

Responses to Dosage 48

Aboiement	Peg
Malingering	Span Line
1.9	Harvest
Females	Column Head
Abevacuation	Peg
*Omits diagnoses . . .	Footnote
Category	Category Head
Inability to walk . . .	Peg (extension)
Frequency of Complaint	Bracer Head

Responses to Dosage 50

(1) Pathogenesis of a Sudden Death
(2) Improved Diagnosis of Myocardial Ischemia by Adding Postexercise Left Ventricular Ejection Time
(3) Hemodynamics During Rest and Bicycle Exercise in Patients with Coronary Artery Disease
(4) Reducing Mortality of Myocardial Infarction Within Six Months
(5) Sudden Death Risk Factors in Coronary Artery Disease

Responses to Dosage 51

Set 1. Delete: Fourteen rabbits were fed . . . .
(Belongs in the Methods section)

Rabbits fed a hypercholesterolemic . . . .
(Belongs in Results)

Set 2. Delete: The terminal now being used with the system . . . .
If the material is well designed it tends . . . .

Responses to Dosage 52

The sequence of information given in the published Method section was as follows: 3, 2, 9, 13, 4, 12, 1, 11, 10, 5, 6, 8, 7.

Responses to Dosage 61

(1) We were not able to discern the meaning of this paragraph, and therefore could not adequately edit it. We solicited a revision from a third person. Although the version is perhaps offensively flip, that flipness shows the reader's frustration at trying to derive meaning from the paragraph. Nevertheless, the revision seems to be a rather accurate interpretation of what was said in the original–though the idea is not otherwise improved.

Revision: People with organic brain disturbances always get depressed. Depressed people, therefore, have organic brain disturbances, I think. I betcha maybe three or four people, even, get these brain disorders. You know, brain disorders are bad because they cause depression (sometimes).

(2) Capuchin Monkeys (*C apella*)–We were only able to obtain two female adults and eighteen juveniles of this species; none had aortic or coronary artery lesions.

(3) REPORT OF A CASE

~~The patient is~~ a seventeen-year-old ~~male, who~~ was noted to have "blue eyes" as an infant ~~by his parents. The patient~~ suffered several fractures ~~as a child,~~ and X-rays taken at each occasion ~~were consistent with the diagnosis~~ of osteogenesis imperfecta. X-rays taken in 1965, 1966, and 1967 showed the characteristic changes of osteogenesis imperfecta: narrow diaphysis, widening of the metaphyses, underdevelopment of the epiphyses, ~~of the radius and ulna,~~ and healed fractures.

Because of ~~the~~ his~~tory of~~ frequent fractures, the child was told to avoid trauma ~~and it was recommended that he~~ not attend public schools. A pes planus deformity was treated ~~with~~

elevation of the inner border of both heels and soles with rubber and leather supports. The patient sustained a fracture to a metacarpal in February, 1971, which healed without sequelae. This is the last known fracture.

In June, 1974, the patient was seen at the Earl K. Long Hospital dermatology clinic with a six month history of multiple asymptomatic firm, erythematous papules, 3-5 mm in diameter, located at the left lateral aspect of the neck and symmetrically distributed on the anterior-lateral aspects of each elbow. A shave biopsy was done, and an excisional biopsy followed one week later. Local therapy with liquid nitrogen and intralesional triamcinolone 5 mg/cc was begun after the diagnosis was made. There has been some flattening of the lesions. All laboratory work has been within normal limits: bilirubin (total), 1.3 mg%; alkaline phosphatase, 75 I.U.; sgpt, 10 I.U.; LDH, 67 I.U.; sodium, 144 mEq; potassium, 5.0 mEq/l; fasting blood sugar, 104 mg%; bun = 15 mg%; hemoglobin, 14.9 grams; and white blood count = 8,300. The patient's electrocardiogram was entirely without abnormalities, the chest x-ray showed no unusual changes, and the patient had a normal waking and sleeping-state electroencephalogram. The VDRL was non-reactive, and fungus studies have been negative.

A chromosome study was performed. The number in peripheral leukocytes was 46 and analysis of several metaphases revealed no structural abnormality of the chromosomes.

The physical exam revealed no abnormalities other than the presence of light blue sclerae and the previously mentioned skin changes. *He* ~~The patient~~ was *slender but* ~~noted to be~~ of average height for his age. ~~but was slender build~~. *He had* ~~There was~~ no hyperelasticity of the skin or articular hyerlaxity. ~~There was~~ no decreased *in* auditory acuity *was* noted, but *no* ~~an~~ audiogram was *taken.* ~~not performed.~~

The patient has been in good health since he first was seen at the hospital. *Although he* ~~The patient~~ has continued to wear corrective shoes for his pes planus deformity, *he leads a rather normal life. He* ~~The patient~~ now attends public schools, and ~~leads a rather normal life.~~ His grades have been good. ~~The patient is one of four children.~~ His *parents and his* two brothers and one sister are in good health and show no evidence of *EPS or osteogenesis imperfecta.* ~~being affected with either of the diseases under discussion. The patient's mother and father are alive and in good health, and are not affected with EPS or osteogenesis imperfecta~~.

Response to Dosage 63

Our interpretation of the Gobbledygook excerpt is—The preceding sentence does not apply to the period of January 1, 1974, through December 31, 1977.

—but, we are not sure we are right!

Is Audit Relevant to the Medical Wards of a Teaching Hospital?

Peter E. Schrag, MD, William A. Bauman, MD

• The initial purpose of an experimental program of close personalized supervision of ward patients by attending physicians in addition to house staff in a teaching hospital was to determine whether such personalized care would raise the quality of care and lower the cost and length of stay. By subjective measurements, an improvement in the quality of care was perceived. However, neither length of stay nor hospital charges were reduced when compared to a control population. Moreover, the data suggest that auditing the length of stay and hospital charges does not necessarily measure the quality of medical care.

(*Arch Intern Med* 136:77-80, 1976)

Controlling the cost while maintaining high standards in the area of medical care has become a national concern. The federal government, third party reimbursement agencies, and organized medicine are in the process of studying various forms of audit and review of professional medical services. A description of Professional Standards Review Organizations can be found in Public Law 92-603 (92nd Congress). Their alleged purpose is to ensure that the medical services for which payment is provided by the Social Security Administration will conform to medical standards. In teaching hospitals, however, the audit and review of professional services has been considered hitherto to be the responsibility of those who are giving supervised instruction. Moreover, a more quantitative, formal audit of professional services rendered in a teaching setting has not yet been developed. In fact, the usefulness of such an audit is still to be demonstrated.[1]

An experimental program, which was an attempt to provide close supervision of house staff in a teaching hospital setting, was conducted on the medical service of Presbyterian Hospital during a two-year period. Its purpose was to determine whether close supervision of the house staff by senior attending personnel could lower the cost of the patients' medical care without seriously compromising the learning experience of house staff. In addition, it was an attempt to provide a system of "one-class" care for a population of patients who were undergoing extensive and potentially dangerous diagnostic evaluations and who, therefore, deserved close supervision by attending personnel. This experiment in house staff supervision also offered an opportunity to test the premise that a medical audit is useful to evaluate the appropriateness of care and effort that has been provided.

METHOD

The traditional Presbyterian Hospital medical ward is staffed by a first-year resident who supervises three interns. These physicians are responsible for approximately 30 patient beds and are, in turn, supervised by two medical staff members who make attending rounds each morning for two hours and are available theoretically at other hours for consultation. Attending physicians make rounds daily for a one-month period. In addition, specialized consultation groups are readily available.

One of the three identical wards was selected arbitrarily to serve as an experimental service. Three attending physicians were assigned to accept ultimate responsibility for the care of patients admitted to the experimental service. Patient admissions to the service were systematically alternated among the three medical wards; there was no preselection of the experimental service patients. Each experimental service attending physician was assigned to work with one of the three interns, while all attending physicians worked with the first-year resident. The attending physician accepted all patients cared for by his intern, thus supervising approximately one third of all service patients at any time. One attending physician was on call every third night and ev-

Received for publication Oct 22, 1974; accepted June 6, 1975.

From the departments of medicine and pediatrics, Columbia University College of Physicians and Surgeons, and the Presbyterian Hospital, New York.

Reprint requests to Department of Medicine, Columbia University College of Physicians and Surgeons, 161 Fort Washington Ave, New York, NY 10032 (Dr Schrag).

Fig. 23-1. Reprint of a model article in the IMRAD format.

Table 1.—Length of Stay and Charges by Service

Subjects	No.		Length of Stay, Days ± SE		Charges/Admission, Dollars ± SE		Mean Age, yr	
	Fall	Spring	Fall	Spring	Fall	Spring	Fall	Spring
Men								
Experimental service	59	41	11.1 ± 0.86	13.1 ± 0.97	1,869 ± 141	2,470 ± 130	56.1	54.5
Control services	128	112	12.1 ± 0.60	11.8 ± 0.60	2,078 ± 111	2,545 ± 138	56.9	51.5
Women								
Experimental service	57	61	13.3 ± 0.97	13.2 ± 0.81	2,330 ± 202	2,586 ± 144	54.2	56.5
Control services	112	124	12.2 ± 0.65	12.7 ± 0.60	2,045 ± 207	2,403 ± 128	58.0	54.3

Table 2.—Prevalence of Disease Categories by Service

Disease Category	Test Period	Men, % Experimental Service	Men, % Control Services	Women, % Experimental Service	Women, % Control Services
Cardiovascular disease*	Fall	30	30	32	39
	Spring	37	27	30	34
Acute myocardial infarction	Fall	7	8	4	5
	Spring	8	10	10	5
Pulmonary disease	Fall	18	22	15	21
	Spring	11	22	16	18
Gastrointestin-tract disease	Fall	26	26	17	14
	Spring	18	22	13	14
Renal disease	Fall	0	0	4	0
	Spring	3	5	0	4
Miscellaneous disease	Fall	19	14	28	22
	Spring	23	14	31	24
Approximate percent	Fall	100	100	100	101
	Spring	99	100	101	99
No. of patients	Fall	59	128	57	112
	Spring	41	112	61	124

* Excluding myocardial infarction.

ery third weekend. All patients were informed that the services of an attending physician were available, accepted this in writing, and were billed for professional services. One of the three responsible attending physicians was always present on morning rounds.

The three physicians who accepted responsibility for the experimental service were employed on a salary (plus incentive when earned) and were able, in addition to their service responsibilities, to maintain private medical practices. The income from both activities was pooled into a general fund from which salaries, expenses, and incentive pay were paid out. After 18 and 24 months of operation, the experimental service was compared to the traditional medical ward services at Presbyterian Hospital in the following respects: (1) nature of the problems that were cared for; (2) length of patient stay; (3) average hospital charge per patient; and (4) number of major diagnostic procedures.

The length of stay and cost data were obtained through the hospital's computerized billing system. Two pilot surveys of data on patients in the fall of 1972 and in January 1973 suggested that meaningful differences in charges and length of stay between the experimental and control services occurred. These results suggested that with private type of care, charges were substantially lower than with the traditional teaching ward setting.

To confirm these results, a more rigorous analysis that covered all of the patients discharged during two later time periods in 1973 was undertaken. Detailed financial data from the hospital computer system provided a complete recapitulation of each patient's length of stay and an itemization of all charges associated with room and board, laboratory tests, and other billable items. The records analyzed were those that belonged to patients discharged during the periods April 1 through June 30, 1973 (spring) and September 15 through December 14, 1973 (fall). These specific periods were chosen to test the differences between care rendered by experienced house officers in the spring and the care rendered by relatively new house staff members in the fall. Bills from 796 discharged patients were retrieved. Bills of patients hospitalized for less than one day or for more than 31 days were excluded from analysis, because these patients had problems that were not appreciably altered by type of care. For the spring of 1973, bills for 356 discharged patients qualified for review, and 338 bills qualified for the fall of 1973. The medical records that corresponded to each of these bills were requested from the medical records department; all but 28 records were available for review.

A college student was taught to abstract the bills and the medical records for length of stay, total charges, diagnoses, sex, age, numbers and types of x-ray films, biopsies, and arteriograms. Most patients had multiple diagnoses. For purposes of analysis, discharge diagnoses were classified by one of us (P. E. S.) into one of the following six problem areas: (1) cardiovascular disease; (2) acute myocardial infarction; (3) pulmonary disease; (4) gastrointestinal disease; (5) renal disease; and (6) other miscellaneous diseases. The abstracts of patients' records consisted of length of stay,

Fig. 23-1. *(Continued)*

total charges, laboratory tests, and diagnoses. They were grouped according to the experimental and control services for the spring and for the fall. Length of stay and charges are tabulated in Table 1. Table 2 compares the distribution of problem areas among each of the groups.

The three attending physicians, who were responsible for care on the experimental service, recorded for each patient admitted to the service what, if any, important contribution to the care of the patient had been made by their intervention. House staff members were also surveyed by anonymous questionnaire as to the helpfulness of the intervention of the attending physician on the experimental service. These evaluations were analyzed and were found to verify that six general aspects of care were improved on the experimental service. Whether the improvement in these six aspects of patient care corresponded to any change in average costs or length of stay was also examined.

RESULTS

There were no meaningful differences in the average length of stay for patients in the experimental service, as compared to the control services, either in the spring or in the fall, and there were no differences in the average length of stay in the spring as opposed to the fall (Table 1).

The average charges per admission also were not meaningfully different between the experimental and control services in either spring or fall. The differences in average charges between the spring and the fall for patients in each of the groups reflected the increase in room, board, and ancillary charges instituted by the hospital on Sept 15, 1973 (Table 1). In all other measured aspects, the patients on the experimental service did not differ from patients on the control services either in the spring or in the fall.

For example, there were no substantial differences in the prevalence of disease categories in the control or experimental services within or between the two seasons of the year (Table 2).

While the audit of cost and length of stay failed to disclose important differences between the experimental and the control services, attending physician intervention on the experimental service was helpful in the following six general aspects of patient care: (1) Unnecessary invasive diagnostic procedures were discouraged and the scope of peripheral diagnostic evaluations was limited. We frequently discovered house staff plans to perform cardiac catheterization to evaluate the nature of heart murmurs in the absence of meaningful cardiac functional impairment. Similarly, invasive diagnostic procedures to evaluate abdominal complaints could be discouraged if laparotomy appeared mandatory in any event.

(2) The long-term effects of chronic illness were estimated earlier, with more rapid preparation for continuing care. House staff were often reluctant to concede the ineffectiveness of therapy and, therefore, that permanent disability would remain and require consideration.

(3) The dying patient was protected from fruitless intensification of care. House staff with little clinical experience tended to overestimate the possible results of immediate therapeutic intervention in the face of long-standing debilitating illness. Advice from attending physicians forestalled heroic but clearly fruitless attempts to treat intercurrent fulminant illness such as acute renal failure and massive gastrointestinal tract bleeding in patients with profound chronic respiratory and cardiac insufficiency or with widespread neoplasia. For house staff to accept the inevitability of chronic disability or death in a particular patient, the relatively limited discussions held during formal morning rounds often did not suffice.

(4) Consultations with other services were expedited. The urgency or appropriateness of consultations or both requested by house staff were often questioned by specialty groups. House staff appeared to seek the help of specialists too early and responded to a specialty group's suggestions without tempering recommendations in accord with their own unique knowledge of the patient's special characteristics. Some guidance in the use of consultations with specialists seemed necessary to make efficient and appropriate use of consultants.

(5) In emergency consultation for the critically ill, traditional attending physician rounds tended to limit house staff-attending physician contact to two hours each morning. The round-the-clock availability of attending physicians was of major help in ferreting out diagnostic and therapeutic avenues in emergency situations when senior help would not have been obtained otherwise.

(6) Important management decisions that were made later in the course were influenced. The traditional attending physician rounds of necessity concentrated on the more recent admissions or the more severely ill patients. Decisions made later in a patient's hospitalization were often made by house staff without being brought up for discussion during formal morning rounds. The assumption that supervision of patient care by attending physicians is required mainly in the earliest part of the patient's hospital stay was not correct necessarily.

Intervention by attending physicians was often thought by both patients and physicians to be of critical importance in the management of individual cases. Nevertheless, such intervention had no demonstrable effect on the average cost or length of stay.

COMMENT

A medical audit procedure was used to determine whether close supervision of house staff had a measurable effect on patient care. Such retrospective chart analysis is the procedure recommended by the Congress (Public Law 92-603) to test the appropriateness of medical services. The analysis is intended to identify variations from standard patient care practices. For this experiment, the audit failed to disclose expected differences in average length of stay, charges, or number of procedures when the experimental, closely supervised service was compared to the standard "teaching" service. Furthermore, the audit did not disclose differences that might be attributed to new house officers as distinguished from others with more experience.

One interpretation of the audit's failure to disclose measurable differ-

Fig. 23-1. *(Continued)*

ences between the closely supervised service and the control service is that it accurately reflected the possibly superfluous nature of personalized supervision in the management of these patients.

Personalized supervision by attending physicians in addition to house staff may have little effect on factors measurable by auditing procedures, because these factors are predetermined already by uniformity of house staff training. However, the inability to demonstrate by an audit the importance of the intervention of an attending physician contrasts with the acknowledged critical importance of intervention in individual cases. Moreover, the usefulness of measuring the average length of stay and hospital charges as a means of protecting the public and ensuring the quality of health services seems questionable if average length of stay and hospital charges in teaching hospitals reflect uniformity of training and a cautious approach to the management of multisystem disease rather than critical intervention by the attending physician in patient management.

An alternate interpretation of the failure of this audit procedure to disclose differences caused by types of management is that the audit was not sufficiently sensitive. A more sensitive audit might attempt to classify patients in a manner that takes into account the complexity of the illnesses and the number of organ systems that are malfunctioning. A more sensitive audit might measure also a patient's satisfaction with close personalized supervision. However, an audit that is sensitive to the variety and complexity of illnesses on a medical service and to the benefits of a patient's satisfaction is difficult to design and costly to perform. Additionally, an audit cannot demonstrate the benefits and value of providing one-class care to a population that deserves expert medical attention.

In conclusion, close personalized supervision of patients by attending physicians in addition to house staff in a teaching hospital setting did not reduce average charges or shorten the hospital stays of these patients. Personalized care may not affect hospital charges or length of stay, because standards of care are already predetermined by uniformity of house staff training. This suggests that auditing the length of stay and hospital charges will not be helpful and that auditing procedures for teaching hospitals have yet to be developed. Such auditing procedures will have to deal with as yet undefined variables that are difficult and costly to measure.

Charles A. Ragan, Jr, MD, Michael H. Cohen, MD, and Jeffrey A. Stein, MD, assisted in the preparation of this study. John W. Fertig, PhD, assisted in the statistical analysis. William C. Bauman abstracted the records.

Reference

1. Kavet J, Luft HS: The implications of the PSRO legislation for the teaching hospital sector. *J Med Educ* 49:321-330, 1974.

Fig. 23-1. *(Continued)*

Index

Index